SpringerBriefs in Space Development

Series Editor

Joseph N. Pelton

For further volumes:
http://www.springer.com/series/10058

Giovanni F. Bignami · Andrea Sommariva

A Scenario for Interstellar Exploration and Its Financing

 Springer

Giovanni F. Bignami
IASF-INAF
Milan
Italy

Andrea Sommariva
META (Beijing) Consulting Company Ltd
Beijing
People's Republic of China

ISSN 2191-8171   ISSN 2191-818X   (electronic)
ISBN 978-88-470-5336-6   ISBN 978-88-470-5337-3   (eBook)
DOI 10.1007/978-88-470-5337-3
Springer Milan Heidelberg New York Dordrecht London

Library of Congress Control Number: 2013931207

Printed on acid-free paper

Springer is part of Springer Science+Business Media (www.springer.com)

Preface

This book is a contribution of an astrophysicist and an economist to the debate on space exploration. Space exploration is an important subject and one of the most contested and in need of careful investigation. We strongly believe that the analysis should go beyond the scientific and technical aspects, and should be enlarged to economics, sociology, political science, history, and cultural studies. They are equally important in understanding what space exploration is and how space exploration can contribute to the progress of humanity.

Space exploration is quintessentially a human endeavor. Migrations and voyages of exploration and discovery have characterized and distinguished humanity since the dawn of its history. Humanity is a wandering race. Throughout the history of homo sapiens, motivations for its wandering have been various: climate instability, pursuit of trade, of power and personal wealth, and of knowledge. Whatever the motivations, the thirst for discovery and exploration seems engrained in our mind. There is no doubt that space exploration is the next big step for humanity. In the opening hearings before the Special Committee on the NASA bill on May 6 1958, its Chairman Lyndon Johnson stated:

> Space affects all of us and all that we do, in our private lives, in our business, in our education, and in our Government. [...] We shall succeed or fail [depending on] our [...] success at incorporating the exploration and utilization of space into all aspects of our society and the enrichment of all phases of our life on this Earth.

In writing this book, we explored the conditions for advancement in manned space exploration and colonization. Our analysis, based on the present knowledge of science and technology, indicates that space exploration and colonization of nearby stars systems could be realized within the next 100 years. However, science and technology define the boundary of what is possible. The realization of manned space exploration and colonization depends on our level of civilization. We strongly believe that space exploration on the scale proposed in this book can only be done on a global cooperative basis.

We are far from that level of civilization. There are many grounds for pessimism on whether we can reach this advanced level of civilization in the near future. Despite all the pessimism, our view is that the situation is not entirely hopeless. The human race is the only one on Earth in creating its own problems; hence it is within its own power to solve them. As Dr. Johnston once said:

> Life affords no higher pleasure than that of surmounting difficulties, passing from one step of success to another, forming new wishes and seeing them gratified.

The authors want to thank Federica Abaterusso, Amedeo Luttwak, Federico Romanelli, Laura Simei, and Paolo Sommariva for reading and commenting on an earlier draft of this book. Their comments are sincerely appreciated.

Giovanni F. Bignami
Andrea Sommariva

Contents

Chapter 1
Introduction

The finer part of mankind will, in all likelihood, never perish.
And so there is no end to life, to intellect and the perfection of
humanity. Its progress is everlasting.

Konstantin E. Tsiolkovsky

Travelling to the stars has always fascinated mankind. It has been the subject of numerous literary works and science fiction stories. When we grew up during the sixties, those objectives seemed achievable. The launch of the Sputnik 1 in October 1957 and the first landing on moon by the American Apollo program in July 1969 created an atmosphere of optimism pervading not only science and technology, but also arts, as in the movie of Kubrick *"2001: A Space Odyssey"*, which explored man destiny in space and the encounter with an alien life form: the black cube.

After the optimism of the sixties, space exploration has entered a state of flux. Despite some spectacular accomplishments in the exploration of the solar system[1] and in the field of science,[2] man exploration of space has been limited to suborbital flights and to their short term permanence in the International Space Station. Many observers think that man does not belong to space and that whatever resources are dedicated to space programs should be directed to scientific research. They are loosing sight that space exploration is not just for scientific research, but also for other goals—economic, commercial, cultural, and in the very long term the survival of the human race—and that man has a leading role in it.

It seems to us that we have lost the sense of direction and the drive that prevailed in the sixties. The question is whether the optimism of the sixties was just juvenile exuberance, which was tempered by growing older, and space exploration does really belong to a very distant future. These are the questions that this book tries to answer.

---

[1] Such as the Mars Rover program, the Pioneers, Voyagers and the Galileo missions of the outer planets of the solar system.

[2] Such as the launch of the Hubble Space Telescope and its astronomical observations.

G. F. Bignami and A. Sommariva, *A Scenario for Interstellar Exploration and Its Financing*, SpringerBriefs in Space Development, DOI: 10.1007/978-88-470-5337-3_1,

Given the enormous distances among the stars,[3] some observers consider interstellar travel practically impossible in the foreseeable future, although theoretically feasible. Impossible is however a relative term. It depends on the knowledge or lack of it of basic laws of physics. In the second half of the 20th century, progresses in the controlled release of nuclear energy, the discovery of laser, the production of antimatter via particle accelerators, and the controlled long range rocketry led by von Braun and co-workers have open the door to possible interstellar exploration.

This book looks at the likelihood of interstellar exploration by developing a scenario. A scenario is not forecasts but a way of understanding the dynamics shaping the future by identifying the primary "driving forces" at work in the present. Moreover, it allows for qualitative changes that are not considered in the quantitative extrapolations of past trends. This is particularly important when analyzing technical progress, as very frequently forecasts of these events are made obsolete by unpredictable innovations and scientific breakthroughs.

The present scenario will examine: (1) the motivations for space colonization; (2) where to go: the search for habitable planets; (3) the present state and prospects for interstellar travel technologies; (4) the financial mechanism to fund such enterprises; and (5) political and cultural issues. In Chap. 2, we turn to the motivations. We believe that the discovery of a habitable planet outside the solar system and the possibility that Earth will no longer define the limit of growth will constitute the strongest motivation for interstellar exploration and colonization.

The key focus of Chap. 3 is the search for extra solar planets, particularly habitable planets. It shows that this search is well advanced and that the probability of finding, in the next 10 years, a habitable planet within 5 to 15 light years from Earth is high. Chapter 4 is about the present state and prospects of the technology for spaceship propulsions capable to reach a significant fraction of the speed of light. Several new and promising concepts are being investigated to provide for fast space travel, including nuclear fusion, laser driven sail, and antimatter.

Chapter 5 will focus on the financing of the exploration program. The initial phase of research and designing should be funded by a mix of public and private money via organizations capable of fostering creativity and innovation. The second phase refers to the construction of transportation infrastructure (space-port) and automated probe exploration for gathering information on the target planet. It has to be financed by public money. We advance the hypothesis that public funding could come out of a reduction of 1.5 to 2 % of global military expenditures. The third phase of manned interstellar exploration and colonization should be funded by private money, thereby optimizing the resources to finance such ventures.

Chapter 6 will focus on the uncertainties surrounding the present scenario. Critical uncertainties are identified as political and cultural. Future interstellar

---

[3] The two Voyagers interstellar spaceships—the fastest ever launched from Earth—are travelling at one-thousandth speed of light. They would need about 70,000 years to go to the nearest star, making human interstellar travel unrealistic.

exploration and colonization requires a civilization in which human beings see themselves as inhabitants of a single planet and in which global governance is conducted on a cooperative basis. This would allow a reduction of global military expenditures that can be channelled to the funding of an interstellar exploration program.

There is no doubt that the substitution of expenditures on instruments of death with peaceful space investments is a win-win situation. Human behaviour is however influenced not only by rational motivations, but also by cultural issues, religious taboos, greed and other less pleasant impulses, in essence our value system. It is not the scope of this book to deal with these issues and to suggest solutions. However, we hope that this book will help in bringing science and scholars of humanities together in challenging a number of conventional social, political, religious and philosophical issues, facilitating the discovery of ways to form a global society less confrontational and based on mutual respect and non violence, which may lead humanity to reach the stars in a not so distant future.

Chapter 2
Motivations for Space Explorations and Colonization

Earth is the cradle of humanity, but one cannot live in a cradle forever.

Konstantin E. Tsiolkovsky

This chapter will try to answer the question of what are the most important drivers leading to interstellar travel by humans within the next 100 years. Some observers claim that the main reasons behind space exploration are: (1) survival of the human species, (2) access to additional natural resources, and (3) the spreading of life in the universe.

Physicist Stephen Hawking once said:

The long-term survival of the human race is at risk as long as it is confined to a single planet. Sooner or later, disasters such as an asteroid collision or nuclear war could wipe us all out. But once we spread out into space and establish independent colonies, our future should be safe. As there isn't anywhere likes the Earth in the solar system, so we would have to go to another star.

We believe that the discovery of a habitable planet/s within 5–15 light years from Earth will be the crucial driver for interstellar exploration and colonization. The survival of the human species is a very long term threat and, as such, may not be the main driver for interstellar exploration within the time horizon of our scenario. However, our analysis indicates that the effects of population growth and limited resources combined with those of "man-made disaster", although not catastrophic, could affect negatively a large part of the world population. This would make the discovery of a habitable planet/s and the possibility of interstellar exploration and colonization even more attractive in the eyes of the general public.

In the second half of the 20th century, thanks to progress in telescope technology, both on the ground and in space, astronomers have detected extra solar planets orbiting distant suns. Over the next decade, space-based constellations of telescopes will enable astronomers to identify Earth-like planets orbiting distant stars. These instruments will also allow analyzing the spectrum of planets' light to ascertain the presence of chemical elements, such as oxygen, necessary to support life.

G. F. Bignami and A. Sommariva, *A Scenario for Interstellar Exploration and Its Financing*, SpringerBriefs in Space Development, DOI: 10.1007/978-88-470-5337-3_2, © The Author(s) 2013

If the existence of a habitable planet capable of sustaining life is confirmed by astronomical observations, interstellar exploration and colonization will raise strong interests at all levels of society, and could induce a change of heart of governments and the private sector in funding programs for the development of interstellar propulsion technologies and for building an orbital space-port for the assembling and launching of spaceships. Robotic space exploration will still continue to play a crucial role in the early phase of space exploration to achieve orbital surveillance capabilities of the targeted planets. This should provide the necessary information for the second phase of manned exploration and colonization.

2.1 Survival of Human Species

Threats to the survival of the human species come mainly from three sources: (1) pressure of population growth on limited resources, (2) human engineered disasters, and (3) natural disaster such as an asteroid collision.

For a better understanding of the complex relationships between point (1) and (2), we can draw on historical experience here on Earth. Although historical comparisons have to be treated with caution, analysis of past events may shed some light on the future, not in detail, but in repeated patterns.

As Mark Twain once said:

> History does not repeat itself, but it rhymes.

One particular migration and colonization is relevant for the present scenario: the Polynesian population of the Pacific Ocean. Modern historians[1] of the Polynesian people before European contact indicate that this journey stretched from 3000 BC to about 800 AD. About 3000 BC speakers of the Austronesian languages, probably on the island of Taiwan, mastered the art of long-distance canoe travel and spread themselves and their languages south to the Philippines and Indonesia, and east to the islands of Micronesia and Melanesia (Fig. 2.1).

They then branched off and occupied Polynesia to the east. Dates and routes are uncertain, but they seem to have started from the Bismarck Archipelago, and then they went west past Fiji to Samoa and Tonga about 1500 BC. By 100 AD they were in the Marquesas Islands and 300–800 AD in Tahiti. The dates 300–800 AD are also given for their arrival at Easter Island, their easternmost point, and the same date range for Hawaii, which is far to the north and distant from other islands.

It is evident that these voyages were not made by chance. Transfer of many species of crops and livestock all over the Pacific islands prove that the settlements were done by colonists carrying products and animals from their homeland, deemed essential for the survival of the new colonies. There is still a debate on the motivations for these migrations. Several arguments have been put forward for explaining these migrations.

---

[1] See Kirk 2000.

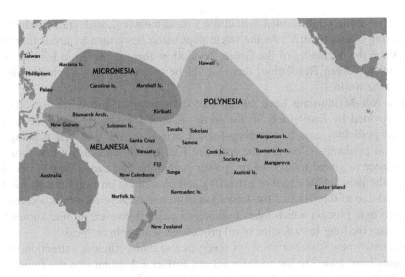

Fig. 2.1 Polynesian people migrations

One refers to population growth in the new settlement. As population grew in the new settlement, it may have put pressures on the limited agricultural resources of the island, inducing a group of colonists to move either into the unknown or to new island previously discovered by voyagers sailing upwind on a predetermined bearing. Another argument is that human activities, as deforestation in the new settlement, posed pressure on a fragile ecosystem thus creating a crisis that could only be solved by migration. In reality, it was probably a combination of these factors that was behind the spreading of the Polynesian people throughout the Pacific Ocean.

2.1.1 Pressure from Population Growth and Limited Resources

At present, human beings have populated every habitable corner of Earth, which has become a single habitat. Pressures from population growth and limited resources are real today. It is conceivable that in the next 50 years the population of Earth will continue to increase. Today world's population is about 7 billion. Although it is difficult to forecast the growth in world population, most observers agree that the world population could hit 9 billion in 2050.

There is still a debate whether the forecasted level of population on Earth will reach the threshold compatible with limited resources. There are two schools of thoughts. The Neo-Malthusianism school predicts that population growth will out-run food supply. The dire predictions of Neo-Malthusians are vigorously challenged by a number of economists. They point out that from 1950 to 1984 the

Green Revolution transformed agriculture around the world, with grain production increasing by over 250 %. As the world population has grown by about four billion since the beginning of the Green Revolution, there is reason to believe that, without the Green Revolution, there would be greater famine and malnutrition around the world.

The Neo-Malthusians have replied that the energy for the Green Revolution was provided by fossil fuels, in the form of natural gas-derived fertilizers, oil-derived pesticides, and hydrocarbon-fuelled irrigation. The potential peaking of world oil production may test the critics of Neo-Malthusians, as oil is of crucial importance to global transportation, power generation and agriculture. Oil depletion is the inescapable result of extracting and consuming oil faster than it can be replaced due to the fact that the formation of new natural petroleum is a continuous geologic process which takes millions of years. However, no one knows for sure when the long-term decline of oil production will begin or "peak".

Recently, new discoveries of oil reserves and more efficient extraction technologies have added to the world reserves, further extending into the future the decline in oil production. The International Energy Agency and the oil industry are convinced that oil production could increase to match demand in the longer term. The continuing depletion of low cost resources, the increased costs of developing new resources, and constraints to access to resources are fundamental factors that will continue to affect oil prices. Therefore, in our opinion, Peak Oil in the foreseeable future means not "running out of oil", but "running out of cheap oil".

High levels of oil price affect negatively developing countries, particularly their agricultural sectors, while their negative effects on agriculture in advanced countries are estimated to be less. It takes 1,000 t of water to produce one ton of grain. Although only 17 % of the world's crop land is irrigated, the irrigated lands produce about 40 % of the world total food. Three quarters of the irrigated land is in developing countries, like Pakistan, Nepal, Bangladesh and Northern India. Irrigation is sensitive to changes in oil prices, particularly for small farmers. Hence high levels of oil prices may affect the amount of land irrigated, impacting substantially on both supply and prices of agricultural products.

These negative effects could be partly overcome by the invention and diffusion of yield-enhancing technologies and farm practices. These innovations are directed to raise productivity by increasing resistance to drought, and heat, and by increasing responsiveness to nutrients and moisture. However, in underdeveloped rural areas, people may lack the information or education necessary to take advantage of improved agricultural methods.

Moreover, rural property rights are often poorly defined, and labourers may lack incentives to care about the future productivity of land they farm. Governments could intervene in overcoming these problems by providing information programs, developing new agricultural technologies, and extending property rights. However, budgetary constraints may prevent many governments in developing countries to act, unless substantial aids id injected by international financial institutions.

2.1.2 Man Made Disasters

Aside from oil, there are other factors that could influence the supply of food in the longer term: the effects of climate change, the loss of agricultural land to residential and industrial development, and growing consumer demand in the population centres of China and India, and of other emerging countries. In this section, we concentrate on climate change, because the other factors only tend to reinforce the effects of climate change.

Climate change is caused by the emission of a number of gases, which causes increased heating of the atmosphere, land and seas. Human activities such as the burning of fossil fuels and land clearing have increased the concentration of most greenhouse gases in the atmosphere. Carbon dioxide (CO_2) is the main greenhouse gas, accounting for approximately two-thirds of the increased trapping of heat in the atmosphere. Fossil fuel combustion and deforestation are major sources of CO_2 emissions today. CO_2 emissions have reached unprecedented levels with respect to pre-industrial ones (Fig. 2.2). The physical impacts of climate change are complex and unpredictable, but will include higher global average temperatures, melting of the ice cap (Fig. 2.3), rising sea levels, and increasing scarcity of agricultural land and freshwater.

We examined several studies on the effects of climate change on agricultural production, based on projections of Earth's climate under an atmospheric CO_2 level that is twice current levels. The climate scenarios are derived from popular General Circulation Models. Rough estimates suggest that over the next 50 years or so, climate change per se may be a less serious threat to meeting global food needs than expected. The sensitivity of agriculture to climate change depends on its direction and magnitude. Higher latitude countries, such as Canada, Russia, and

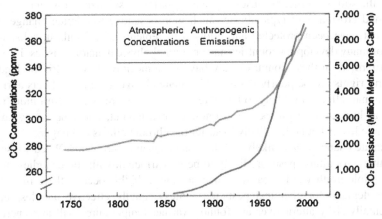

Source: Oak Ridge National Laboratory. Carbon Dioxide Information Analysis Center, http://cdiac.esd.oml.gov/.

Fig. 2.2 Carbon dioxide levels

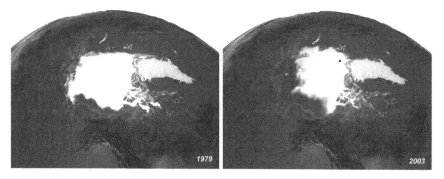

Fig. 2.3 Melting of the ice cap (Aurenkaplan)

most of the advanced countries, are predicted to benefit from a warmer climate, since this could open regions that are currently too cold to farm. Conversely, tropical countries are likely to be losers in a warmer, drier climate. These developments could result in a zero or positive effect on the global agricultural supply.

Although these predictions may be the best available at present, they are subject to great uncertainty. In particular, achieving predicted increases in agricultural supply will require substantial diffusion of agricultural technologies, development of infrastructure, and improvements in irrigation. Much will depend on future government policies to promote investment and to increase efficiency in agriculture. For example, the potential for increasing food production in the former Soviet Union and Eastern European countries is thought to be substantial, if waste during distribution is reduced, property rights are extended, farms are decentralized, and restrictions on food exports and import of technologies are removed.

Climate change will likely aggravate regional agricultural problems. Areas where the soil is fragile, because of intensive farming, salinization, water logging, and wind erosion, will typically be much more sensitive to climate change than soils that have been protected and allowed to replenish. The soils in substantial parts of many developing countries have marginal physical characteristics. Further, tenant farmers in these countries may have little incentive to care for the land as property rights are be poorly defined. If climate change adversely affects agriculture, human effects are likely to be more severe in a poorer world with more people near hunger. In a prosperous economic environment, consumers may suffer economic losses but be less likely to suffer the chronic effects of hunger.

Climate change is also likely to affect other resource problems. Weeds, insects, and other agricultural pests are likely to be redistributed, with some studies showing possible north ward expansion of pest ranges. If this occurs, there may be a greater demand for chemical pesticides unless equally effective and less environmentally risky alternatives are found. Warmer temperatures will also increase water demands for crop growth. In areas where increased water demand is not offset by additional rainfall or irrigation water supplies, climate change may further intensify the competition between growing urban, industrial, recreational, environmental, and agricultural users of water. Again, the effects will be different in various regions of the world. The region most at risk from problems of food security is Sub-Saharan Africa, which is more sensitive to reduced rainfall, rainfall variability, and evaporation than other regions. Only around 2 % of its cropland is irrigated, 50 % is in arid or semi-arid areas, and much of its soil is very fragile.

2.1.3 Geopolitical Implications

In the foreseeable future, neither the dire predictions of the Neo Malthusians in terms of generalized famine and malnutrition nor the dire predictions of the Peak Oil theory are likely to materialize. However, population growth, limited resources and climate change are likely to affect differently various regions of the world, with important geopolitical implications.

The geopolitical implications of climate change are uncertain. It is vastly difficult to forecast how nations will respond to environmental pressures, which are likely to be stronger in the developing world, while advanced countries will be less affected because of their geographical position. Threats to international security are most likely to emerge in countries where governance capacity is overstretched and which unable to manage the physical impacts of climate change. Where this occurs, civil unrest, inter-communal violence, mass migration, breakdown of trade, state failure, and international instability become increasingly probable.

Faced with these uncertainties, one can foresee two scenarios. The worst case scenario is one in which: (1) multilateral negotiations to prevent dangerous climate change collapse; (2) efforts to strengthen the adaptive capacity of the most vulnerable states are blocked; (3) localized impacts may trigger the political and economic collapse of a key pivot state; (4) these localized events will be global in their impact as global networks of trade and diplomacy, which sustain the security and prosperity of many states, are increasingly disrupted; (5) where global processes are too slow, those states worst affected by disruption to global supply chains are increasingly likely to pursue ad-hoc solutions which could trigger further disputes, including interstate conflict.

A more benign scenario is one in which: (1) multilateral negotiations to prevent dangerous climate change continue; (2) efforts to develop clean energy sources, including nuclear fusion, are successful; (3) efforts to strengthen the adaptive capacity of the most vulnerable states continue, but the degree to which this occurs is dictated primarily by the economic resources and the pace of environmental change; (4) localized impacts could potentially trigger the political and economic collapse of already fragile states; (5) these localized events will be global in their impact. However, shared interests make conflict over intervention unlikely; (6) where global governance structures are overstretched, there is likely to be greater reliance on regional organizations working in partnership with traditional global governance structures.

At present, it is unclear which scenario might develop. How nations will respond to climate change depend largely on the extent to which the impacts are felt economically; and how this will be mediated by international financial institutions, access to markets, and the continuation of trade. Although the two scenarios differ on the degree of violence and global instability, it is conceivable that, in the next 50 years, life of a large part of the world population in developing countries would be, in the worlds of Dr. Johnston, "short, nasty and brutish".

Advanced countries will be less affected by climate change because of their geographical positions. However, rising food prices and high price levels of energy would affect consumers in advanced countries with probable negative effects on economic growth and income inequalities. Moreover, mass migrations from less developed countries, whether legal or illegal, will put enormous pressures on the democratic institutions of advanced societies, as manifested already by the emergence and the growing importance of xenophobic sentiments and political parties in our advanced societies.

2.1.4 Natural Disasters

Natural disasters are referred here as the collision of a meteorite, asteroid, comet, or other celestial object with the Earth. A large number of asteroids orbit the inner solar system, mostly between the orbits of Mars and Jupiter, and this region is known as the asteroid belt (Fig. 2.3). Comets have a wide range of orbital periods, ranging from a few years to hundreds of thousands of years. Short-period comets originate in the Kuiper belt, which lie beyond the orbit of Neptune. Longer-period comets are thought to originate in the Oort cloud, a hypothesized spherical cloud of icy bodies in the outer Solar System. Throughout recorded history, hundreds of minor impact events have been reported, with some occurrences causing injuries, property damage or other significant localized consequences. There have also been major impact events throughout Earth's history which severely disrupted the environment and caused mass extinctions (Fig. 2.4).

There are two schemes for classification of impact hazards from non-Earth objects (NEO): (1) the simple Torino Scale; [2] and (2) the more complex Palermo Technical Impact Hazard Scale.[3] We have calculated the frequency in terms of years according to the Palermo Technical Impact Hazard Scale. We have to note that there is a large uncertainty using this formula because of the uncertainties in determining the energies of the atmospheric impacts. Data are reported in Table 2.1.

Small objects frequently collide with the Earth. There is an inverse relationship between the size of the object and the frequency that such objects hit the earth. The lunar crate ring record shows that the frequency of impacts decreases as

---

[2] See Binzel 1997.

[3] See Chesely et al. 2005.

Fig. 2.4 Inner solar system: Mercury, Venus, Earth, Mars and Asteroids (http://solstation.com)

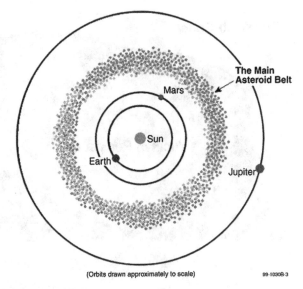

(Orbits drawn approximately to scale)   99-10308-3

Table 2.1 Frequency of collision

Impact in terms of megatons	Frequency in years
10	1,088
50	4,600
100	8,600
500	36,800
1,000	68,700
10,000	545,000
1,000,000	35,000,000

approximately the cube of the resulting crater's diameter, which is on average proportional to the diameter of the impactor. Asteroids with diameters of 5–10 m enter the Earth's atmosphere approximately once per year, with as much energy as 15 kt of TNT. These ordinarily explode in the upper atmosphere, and most or all are vaporized.

Objects with diameters over 50 m, with an energy of about 10 mt, strike the Earth approximately once every thousand years, producing explosions comparable to the one known to have detonated above Tunguska in 1908. Asteroids with a 1 km diameter with energy of about 10,000 mt strike the Earth every 500,000 years on average. At least one known asteroid with a diameter of over 1 km has a possibility of colliding with Earth on 2880. Large collisions happen approximately once every 10 million years. A large impact event is commonly seen as a scenario that would bring about the end of civilization. The last known impact of an object of 10 km or more in diameter was during the Cretaceous-Palaeogene period, which caused the extinction event 65 million years ago (Fig. 2.5).

Until the 1980s this idea was not taken seriously, but opinions changed following the discovery of the Chicxulub Crater. This crater is a prehistoric impact crater buried underneath the Yucatan Peninsula in Mexico. Its centre is located near the town

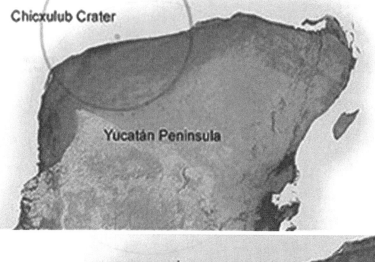

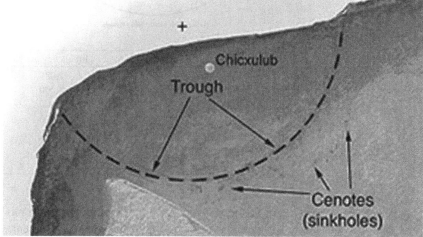

Fig. 2.5 Chicxulub Crater, image from NASA's radar topography mission (NASA, http://photojournal.jpl.nasa.gov/catalog/PIA03377)

of Chicxulub, after which the crater is named. The crater is more than 180 km in diameter, making the feature one of the largest confirmed impact structures on Earth; the impacting bolides that formed the crater was at least 10 km in diameter.

Are we in danger of being erased from the universe? The above analysis indicates that the probabilities of a large impact in the foreseeable future are very low. But there is no question that a cosmic interloper will hit Earth sometime in the more distant future, and we will not have to wait millions of years for it to happen. Objects more than a 1 km wide, impacting Earth every 500,000 years, would touch off firestorms followed by global cooling from dust kicked up by the impact. Humans would likely survive, but civilization might not.

An asteroid 10 km or more wide would cause major extinctions, like the one that may have marked the end of the age of dinosaurs. For a real chill, look to the Kuiper belt, a zone just beyond Neptune that contains roughly 100,000 ice-balls more than 80 km in diameter. The Kuiper belt sends a steady rain of small comets earthward. Sometimes, a bigger one wanders toward us, as witnesses by the comet Shoemaker's impact on Jupiter. If one of the big ones headed right for us, that would be it for pretty much of all higher forms of life, even cockroaches.

In the very long term, there is also the question posed some time ago by Robert Frost when he asked whether the Earth will end in fire or ice. Using the laws of physics, we can reasonably predict how the world will end. Billions of years from now, the sun will gradually expand and consume Earth. We estimate that the sun will heat up by approximately 10 % over the next billion years, scorching the Earth. It will completely consume the Earth in 5 billion years, when the sun will mutate into a red giant. This is a law of physics.

When physicists finally understood how the Sun's alchemy worked following the work of F. Hoyle on a star nucleosynthesis in the early 1950s, this realization caused a substantial impact on the intellectual community. Bertrand Russell then lamented:

> ... that no fire, no heroism, no intensity of thought or feeling can preserve the life beyond the grave; that all the labours of the ages, all the devotion, all the inspiration, all the noon-day brightness of the human genius are destined to extinction in the vast death of the solar system and that the temple of Man's achievements must inevitably be buried beneath the debris of the universe in ruin.

2.2 Key Motivation Factors

We suggest that the main driver for interstellar exploration and colonization is the discovery of a habitable planet capable of supporting life. If in the next decade a habitable planet is discovered, this would change humanity perspectives. As the resources of space will be opened up for use, the Earth will no longer define the limits of growth, which will benefit humanity and not just the few.

Once an Earth-like planet capable of supporting life is found, societies and individuals will be more likely to have a visceral reaction in support of interstellar exploration. *First*, the discovery of a habitable planet would rekindle the interest of the general public for space. Space science and technologies will offer ways to deepen the development of human potentialities. General interest for space would included a set of ideas ranging from exploration as a human imperative and a part of the human experience to a possible evolution of human consciousness that would be concomitant with embarking on interstellar exploration.

Second, the discovery of a habitable planet will certainly raise the interest of wealthy corporations and individuals. In the short term, the development of interstellar exploration technologies would have spill over effects in a variety of fields ranging from new materials, nanotechnology, computer technologies, and communication technologies to new form of energy, particularly nuclear fusion, which will benefit directly these corporations. In the longer term, access to new sources of raw materials and the profits from interstellar commerce would create opportunities for enormous accumulation of wealth, which would certainly stimulate the interest of these corporations and wealthy individuals.

Third, changing perspectives at the general public and corporations levels would certainly facilitate the political process of the allocation of public resources for the development of interstellar exploration. This would mark a strong departure from the past experience of manned space exploration of the 1950s and 1960s, which was driven by a "Space Race" between the Soviet Union and the United States, and was often used as a proxy competition for geopolitical rivalries during the Cold War.

The effects of the discovery of a habitable planet would be comparable, but bigger, than the ones occurred at the time of the discovery of the Americas by Columbus. That discovery revived the exploration instincts of a population, which for centuries had been confined in the narrow spaces of Europe. The discovery of Americas can be seen as a bridge between middle Ages and the Modern era, along with its contemporary Renaissance movement, triggering the early modern period. Accounts from distant lands and maps spread with the help of the new printing press fed the rise of humanism and worldly curiosity, ushering in a new age of scientific and intellectual inquiry. It stimulated the risk taking instincts of the entrepreneurs facing the riches from the new continent, creating the conditions for the revival of the European economies.

There was however a dark side in the colonization of the Americas. The Europeans took land and riches from other people and, in the process, destroyed entire civilizations which had flourished independently. The major difference between now and then would be that we would not go to take land from other "people", as it is unlikely that advanced intelligent life will be found in planets situated at the distance of 5–15 light years away from Earth. There are three arguments here. *First*, all searches for advanced intelligent lives within our galaxy have given no results. Search for advanced extraterrestrial life started in 1960 after the influential paper written in 1959 by Giuseppe Cecconi and Phillip Morrison, who suggested that listening to microwave radiation between 1 and 10 GHz would be the most suitable range to discover artificial signals.

In 1960, astronomer Frank Drake used a radio telescope 25 m in diameter at Green Bank, West Virginia, to examine the stars Tau Ceti and Epsilon Eridani in search for artificial radio signals. He found nothing of great interest. In 1979 the University of California launched a project named SETI. In October 2007, the Allen Telescope Array started to work. It is a "Large Number of Small Dishes" array designed to be highly effective for simultaneous surveys of conventional radio astronomy projects and SETI observations at centimetre wavelengths. Despite great efforts and improved radio telescope technology, artificial radio signals, indicating the presence of advanced intelligent life, have not been identified.

Second, many scientists have tried to calculate the odds of finding intelligent life somewhere out there. They calculate the number of advanced civilizations by an equation known as the Green Bank formula (see Box 2.1). Some of the terms of this equation can be calculated with considerable confidence but others are not, which makes it difficult to draw firm conclusions of any kind. However, even if in the future we can overcome the problem of calculating all the terms of this equation with some degree of confidence, this equation applies to our galaxy that is populated by billion of suns. What are the odds of finding intelligent life in our small corner of the galaxy within 5/15 light-years from Earth? It is evident that the odds are very small if not null.

Third, the search for advanced intelligent life has been based on the argument that a significant fraction of habitable planets would evolve advanced technical civilizations capable at least of using radio signals. This assumes that these life forms are both intelligent and dexterous, which are qualities that are necessary to develop technical tools such as radios. As remarked by some biologists and anthropologists, this argument is based on our own evolutionary experience and assumes that intelligence and dexterity is the best way of evolution. However, as once remarked by Jared Diamond:

> Earth history supports the opposite, as very few animals have bothered with both intelligence and dexterity. Dolphins are smart and spiders are dexterous. Only chimpanzees have acquired some intelligence and dexterity, but they have been rather unsuccessful. Thus our own history indicates that the probability of evolution of a technical advanced civilization at least capable of using radio signals is very low both here on Earth and in our galaxy.

We have not yet said anything about the probabilities of the evolution of other forms of life in extra solar planets. Astronomical observations indicate that the building blocks of life, amino acids, have been found in interstellar clouds and in meteorites. Although we do not know yet how life originated on Earth, there is nothing that prevents the possibility of free living organisms arising in other planets. As extra-terrestrial life goes, simple prokaryotic-like and eukaryotic-like cells are by far the most promising, while complex animal and plant life requires more stringent and stable conditions, like plate tectonics, ocean of water, continental land masses, oxygen-rich atmosphere, and the presence of a large moon.

2.3 A Possible Scenario

It is conceivable that, once a habitable planet capable of supporting life is found, the political will arise to fund the development of propulsion technologies for interstellar exploration. The possible candidates for these technologies capable of travelling at a significant fraction of the speed of light will be examined in Chap. 3 of this book. It is also conceivable that, before a manned expedition, autonomous probes will be sent to the targeted planet/s in order to test the new propulsion technologies and gather the information necessary for the first group of colonists to survive. Only then a first manned expedition would be carried out in order to establish a first colony, which will be a beachhead for later immigration of people from Earth.

Manned exploratory and colonization ventures will be facing a number of problems including problems during the voyage to the planet, and problems for the survival of the colonists in the new planet.

2.3.1 Problems Arising During Interstellar Voyages

Future interstellar space explorers and colonists will face many daunting challenges, including, among others, physiological debilitation, radiation sickness and psychological stress. Likewise ancient mariners faced many challenges in sailing the Pacific Ocean, including: (1) finding out where they were, although they had no difficulty in deciding where they were going[4]; (2) health problems; and (3) psychological problems of living in confined space, while facing the fury of the elements and the unknown. It is a testament to their courage and discipline if they succeeded. It is also an indication that the human race may be more resilient than commonly thought in overcoming obstacles when motivated by survival.

We have several advantages with respect to ancient mariners. *First*, once we have identified a habitable planet and received information from interstellar probes, we will know where to go.

There are many criteria by which to choose the prime targets for an interstellar probe mission. The criteria and the list will change as we search the neighbouring stellar systems for planets. At the present time, there are seven prime targets within 12 light years of the Sun as described in Chap. 3, the closest one being Alpha Centaury, which is 4.3 light years from Earth.

Problems of navigation among the stars have yet to be solved, but are not outside present technologies. A space traveller would get confused and probably get lost among the stars if he used an earth base star chart to navigate. As one travelled deeper into space they would need a computer that could translate a stars known absolute magnitude into its apparent visual magnitude from that position. Visual navigation among the stars would be difficult to say the least. Even with a computer generated chart the navigator would be starring into unfamiliar skies (Fig. 2.6).

Navigation in space would be more likely done by clocks. These clocks would be very accurate time pieces that kept the proper time even at the high speeds necessary to travel through space. The space navigator could then compare the time of a known position with the astronomical time of their current position. These two times would then be used to calculated distance travel and current location.

---

[4] Navigators travelled using only their own senses and knowledge passed by oral tradition. In order to locate directions at various times of day and year, navigators memorized important facts: the motion of specific stars and where they would rise on the horizon of the ocean; weather; times of travel; wildlife species; directions of swells on the ocean, and how the crew would feel their motion; colours of the sea and sky, especially how clouds would cluster at the locations of some islands; and angles for approaching harbours.

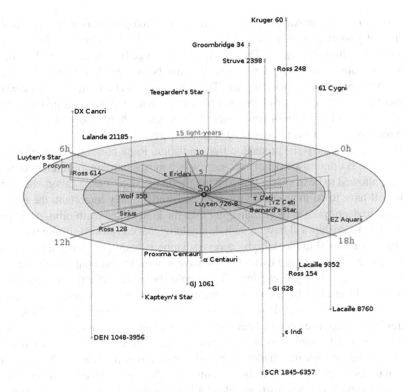

Fig. 2.6 Closest Star systems to Earth (http://en.wilkipedia.org/wiki/List_of_Nearest_Stars)

Astronomical time is the time that is determined by an astronomic object or phenomena. For example, on earth we can tell what time it is from any location by the suns position or a stars position. We would have to find something similar to gage space travel by. Maybe a variable star or a known stars intrinsic brightness could be used. The stars absolute magnitude values would become extremely import to the traveller. Stars could be identified by their absolute magnitudes, spectrum and other physical properties that finger print them.

Second, an advantage of present day with respect to the past is the presence of modern medicine, which will help in devising ways[5] to overcome health and psychological problems faced by explorers and colonists in long term space travel. Ongoing research both at the International Space Station and in laboratory here on Earth will contribute to the solution of these problems for future interstellar explorers and colonists. The main objective is to discover how well and for how long people can survive the extreme conditions in space, and how fast they can adapt to the Earth's environment after returning from their voyage.

---

[5] Some of the problems involved in long term space flights have been analyzed by the HUMEX study as reported by G. Horneck et al. 2003.

Research and technological developments on physiological debilitation due to the absence of gravity is well advanced, although some of this technology has not yet been tested on human subjects. These technologies range from specialized drug mixtures to vibrating machines. A solution[6] that has been recently discovered to treat loss of bone density is a moving plate. The plate vibrates at a frequency of between 20 and 50 Hz. The inventor of this machine has found evidence that this vibrating contribute to the formation of new bones. After a year at 20 min session per day, the bone mass of a sheep has increased 20 % as compared to a control group.

The difficult issue is the protection from radiation. Right now, the countermeasure for radiation is limiting astronaut exposure, which means limiting the amount of time they are allowed to be in space. But on a long-term mission of exploration, the astronauts will have to be in space for years on end, and, equally important, the type of radiation in deep space is more damaging than the kind in low earth orbit. An exploration class spaceship will have to include shielding that can absorb cosmic rays.

The best material to block high-energy radiation is hydrogen. Since it is difficult to make a shield out of pure hydrogen, we have to look for materials than have high hydrogen content, like polyethylene. To completely block radiation, hydrogen-rich shields would need to be a couple of meters thick, which is impractical, because of the weight and volume. However 30–35 % of the radiation can be blocked by shields just five to seven centimetres thick, which might be the most efficient choice.

Astronauts and colonists would still need to cope with about 70 % of the radiation that is getting through the shields. A possible solution is medication. Antioxidants like vitamins C and A can help by sopping up radiation-produced particles before they can do any harm. Scientists are also looking for ways to help the body after the damage has been done. One researcher may have found a way to instruct a damaged, abnormal cell to destroy itself. Another researcher is exploring the cell cycle. As a cell divides, it pauses occasionally, to check its genes for any kind of damage and to repair errors. With pharmaceuticals that lengthen this part of the cycle, researchers believe they can give the cell more of a chance to fix its own problems.

Even if we could prevent the damage caused by radiation and weightlessness that would still be only part of what is need. No matter how healthy and fit astronauts and colonists are, the possibility exists that some medical problem could arise during long missions. Astronauts will need to treat any such illness or accident by themselves, using only the tools they've carried with them. This means developing technologies that are as smart and as capable as possible. It means developing expert systems that can work effectively regardless of the training of the people who are operating them.

Robotically assisted surgery might also play a role. In space, minimally invasive surgery will be important. By using robots, which can make steadier, more even movements that a human hand, surgeons can make smaller, finer incisions than they could on their own. Telemedicine will be another key tool, and that too is already being explored. Other possible technologies abound. Consider a device

---

[6] See Rubin and Mcleod 2002.

that could produce medicines from stored substrates, only when the medicines were needed. Long term exploration missions are likely to exceed the life of many pharmaceuticals. If you could produce pharmaceuticals as you needed them, then healthy life might be much less problematic.

Third, to satisfy human needs on interstellar exploration, the spaceship must be equipped with life support systems capable of recycling as much as the waste products as possible. A life support system capable of recycling the atmosphere and water would reduce the mass of consumables by 90 %. The most promising systems are biological, based on algal and microbial reactors. This is so because of their relatively small size, the possibility of controlling them, and, in case of failure, the possibility of restarting the system relatively fast.

In summary, advances in genetic, biology, immunology, pharmaceutical science, robotics, and intelligent systems would contribute to make interstellar exploration feasible. It is interesting to notice that most of these advances are already taking place here on Earth because of the needs to advance cancer treatments, and improving the efficiency of medicine through telemedicine and robotic, contributing to the feasibility of interstellar exploration.

2.3.2 Problems of Survival

The first wave of future colonists will also face the challenge of surviving in the new environment before the return of the spaceship from Earth. A number of questions have to be answered on this respect.

First, who will be the future explorers and colonists? Today, as yesterday, personal values are largely diversified among the human population. There are many individuals who place high value in individual freedom, personal control and self-realization, and in living free from the constraints imposed by the conditions of overcrowded conglomerates: the modern version of the frontier men and women. A recent survey[7] indicates that these people represent about 25 % of the population in advanced industrial countries. Others place high values on the search of new opportunities or on the simple desire for adventure and exploration, the modern version of Dante's Ulysses. All these individuals would be attracted by the opportunities arising from interstellar space exploration and colonization, and would constitute the bulk of future interstellar space explorers and colonists.

Second, how many people would be necessary for establishing a beachhead colony in a habitable planet? This would depend on the number of years that the beachhead colony will remain isolated before the return of the spaceship from Earth, including the possibility that this period could be very long. Researchers in conservation biology have tended to adopt the "50/500" rule of thumb. This rule says a short-term effective population size of 50 is needed to prevent an unacceptable rate of

---

[7] See Klinenberg 2012.

inbreeding, while a long-term effective population size of 500 is required to maintain overall genetic variability. The population size of 50 prescriptions corresponds to an inbreeding rate of 1 % per generation, approximately half the maximum rate tolerated by domestic animal breeders. The population size of 500 attempts to balance the rate of gain in genetic variation due to mutation with the rate of loss due to genetic drift. We thus assume that the initial number of colonists should be in the range of 50/100, which would allow normal reproduction for about 30/60 generations.

Third, careful selection of the first colonists according to needed expertise and psychological profiles and advanced planning of necessary supplies to bring about will go a long way in increasing the chances of survival of the first colonists. Among the expertise needed by the first colonists is engineering expertise, geological expertise, in order to explore the geology of the planet, biogeochemical, in order to carry out search for life works, agronomical and medical expertise. Other functions such as administrative, organizational and legal, and logistical should be assigned to individual colonists on the basis of their individual cross training. The psychological profile of the first colonists should focus on the "frontier" type, characterized by strong sense of individuality, personal control, and self realization as these first colonists will face a long period of isolation and have to rely on their inner resources to survive.

2.4 Conclusions

Many of the arguments mentioned in this chapter have been present in the works of great science fiction writers of the past. Advances in science in the last 100 years, coupled with the extraordinary development of technologies, are making the speculative groundwork of science fiction writers closer to reality. In the next chapters, we will go in greater details to examine how changes in science and technology will make: (1) probable the discovery of habitable planets outside the solar system; and (2) realistic the development of a propulsion technology capable to reach a significant fraction of the speed of light.

Among all these changes there is a constant: humanity ingrained thirst for exploration. During the early history of our species, migrations out of Africa were most likely the result of the extremely variable climate in the last part of Pleistocene, rather than conscious motivations. At the end of the Pleistocene, homo sapiens had colonized the entire world with the exception of Antarctica.

Voyages of exploration and discovery came later on. The Phoenicians explored all the Mediterranean Sea in search for trade and, in the process, established many colonies. They ventured out to the Atlantic Ocean, north toward Ireland and England, and they circumnavigated Africa. Voyages of exploration culminated in the early 15th century and continued through the 19th century, during which Europeans explored Africa, the Americas, Asia and Oceania.

When one casts his or her mind back in time from the origin of our species in Africa to the colonization of the globe by humanity and imagines how our ancestors survived against all odds and the most extreme environments, one marvels at our ancestors' ingenuity and adaptability. It also comfort us at the time in which the human race may be closing in to the next big migration towards the stars.

Box 2.1
Drake Equation

The Green Bank equation, also known as Drake equation, is a mathematical equation used to estimate the number of detectable extraterrestrial civilizations in the Milky Way galaxy. It is used in the field of the Search for Extraterrestrial Intelligence (SETI). The equation was devised in 1961 by Frank Drake, now Emeritus Professor of Astronomy and Astrophysics at the University of California, Santa Cruz.

The Drake equation states that:

$$N = R^* \times f_p \times n_e \times f_l \times f_i \times f_c \times L.$$

Where

N The number of civilizations in our galaxy with which communication might be possible;

R^* The average rate of star formation per year in our galaxy;

f_p The fraction of those stars that have planets;

n_e The average number of planets that can potentially support life per star that has planets;

f_l The fraction of the above that actually go on to develop life at some point;

f_i The fraction of the above that actually go on to develop intelligent life;

f_c The fraction of civilizations that develop a technology that releases detectable signs of their existence into space;

L The length of time for which such civilizations release detectable signals into space.

Considerable disagreement on the values of most of these parameters exists, but the values used by Drake and his colleagues in 1961 were:

- $R* = 1$/year (1 stars formed per year, on the average over the life of the galaxy; this was regarded as conservative);
- $f_p = 0.2–0.5$ (one fifth to one half of all stars formed will have planets);
- $n_e = 1–5$ (stars with planets will have between 1 and 5 planets capable of developing life);
- $f_l = 1$ (100 % of these planets will develop life);
- $f_i = 1$ (100 % of which will develop intelligent life);
- $f_c = 0.1–0.2$ (10–20 % of which will be able to communicate);
- $L = 1000–100,000,000$ years (which will last somewhere between 1000 and 100,000,000 years).

Drake states that given the uncertainties, the original meeting concluded that $N \approx L$, and there were probably between 1000 and 100,000,000 civilizations in the galaxy.

Criticism of the Drake equation follows mostly from the observation that several terms in the equation are largely or entirely based on conjecture. Thus the equation cannot be used to draw firm conclusions of any kind. One reply to such criticisms is that even though the Drake equation currently involves speculation about unmeasured parameters, it was not meant to be science, but intended as a way to stimulate dialogue on these topics. Then the focus becomes how to proceed experimentally. Indeed, Drake originally formulated the equation merely as an agenda for discussion at the Green Bank conference.

References

Binzel, R.P.: A near-Earth object hazard index. Ann. New York Acad. Sci. **822**, 545–551 (1997)

Chesley, S.R., et al.: The Palermo Technical Impact Hazard Scale. NASA/JPL Near-Earth Object Program Office. http://neo.jpl.nasa.gov/risk/doc/palermo.html19 (2005)

Horneck, G., et al.: HUMEX: A Study on the Survivability and Adaptation of Humans to Long-Duration Exploratory Missions. ESA Publication, Paris (2003)

Kirk, P.V.: On the Road of the Winds: An Archaeological History of the Pacific Islands before European Contact. University of California Press, Berkeley (2000)

Klinenberg, E.: Going Solo: The Extraordinary Rise and Surprising Appeal of Living Alone. Penguin, London (2012)

Rubin, C., Mcleod, K.: Quantity and quality of trabecular bone in the femur are enhanced by a strongly anabolic, noninvasive mechanical intervention. J. Bone Miner. Res. **17**(2), 349 (2002)

Chapter 3
Where to Go: The Search for Habitable Planets

Do there exist many worlds or is there but a single one?
This is one of the most noble and exalted questions in the study of Nature.

Saint Albert the Great

Philosophers and scientists speculated for centuries about the existence of extra solar systems. This speculation was not very common among ancient Greek scientists. Given the scale of the universe generally accepted at their time, the existence of extra solar systems gathered very little support. According to Aristotle and Plato, the universe was a finite sphere. Its ultimate limit was the "primum mobile", whose diurnal rotation was conferred upon it by a transcendental God, not part of the universe, a motionless prime mover and first cause.

Aristarchus estimated the distance from the Earth to the celestial sphere in about one million miles, which is the present day estimate of the distance between Earth and Saturn in our solar system. The notable exception was Democritus. In the decades around 400 BC, he deduced that our Milky Way is composed of stars that appear to be pinpoints of light because of their enormous distances. However, at the time, Democritus intuitions were not taken seriously.

Aristarchus's heliocentric model of the solar system could have led to different conclusions about the size of the universe. One of the criticism of this model centred on the expected displacement of the stars resulting from the movement of Earth around the sun (known as stellar parallax), which was not observed. The contrary argument would have been that the null stellar parallax could be explained by the stars being very far away, hence the possibility of the existence of other solar systems. Since, at that time, there was no way to measure the scale of the universe, Aristarchus's ideas on the movement of Earth were not incorporated in the classical Greek geocentric model of the universe.

The Greek classical geocentric model found its most important systematiser in Ptolemy. Ptolemy's model was enshrined in his "Hé megalé syntaxis" (The Great Collection). This model was almost universally accepted and was later on adopted by the Catholic Church. It gathered wider audience when it was translated in

G. F. Bignami and A. Sommariva, *A Scenario for Interstellar Exploration and Its Financing*, SpringerBriefs in Space Development, DOI: 10.1007/978-88-470-5337-3_3,
© The Author(s) 2013

Arabic and re-titled the Almagest. One had to wait about a 1,000 years before the rediscovery of Aristarchus and Democritus ideas.

In the first half of the 15th century, Nicolaus Cusanus dropped the Aristotelian cosmos. Nicolas Cusanus envisioned instead an infinite universe, whose centre was everywhere and circumference nowhere, with countless rotating stars of equal importance. He also considered that neither were the rotational orbits circular, nor was the movement uniform.

Copernicus' epochal book, "De revolutionibus orbium coelestium" (On the Revolutions of the Celestial Spheres), was published just before his death in 1543. It put again the Sun at the centre of the universe. His work stimulated further scientific investigations by Kepler, Galileo and Newton, becoming a landmark in the history of science that is often referred to as the Copernican Revolution.

However, Copernicus, Kepler and Galileo still thought that the universe was finite and spherical. It was big enough to contain the orbits of the six known planets. It was only Giordano Bruno—an Italian Dominican friar, philosopher, mathematician and astronomer—that put forward the radical view that space is infinite, that the fixed stars are similar to the Sun and are likewise accompanied by planets. His cosmological theories went beyond the Copernican model in proposing that the universe contained an infinite number of inhabited worlds populated by other intelligent beings.

In the 18th century, the possibility of stars similar to the sun and accompanied by planets was mentioned by Isaac Newton in the "General Scholium" that concludes his Principia. Making a comparison to the Sun's planets, he wrote: "And if the fixed stars are the centres of similar systems, they will all be constructed according to a similar design and subject to the dominion of one".

Despite the invention of the telescope and its subsequent improvements, there was no way of knowing how common these star systems were, how similar their planets might be to the planets of the Solar System and what was the scale of the Universe. No matter how brilliant the ideas proposed by these scientists, this was no real physics but simply conjectures.

One had to wait the 20th century and progress in astronomic observations for an answer to those questions. However, these discoveries did not come suddenly, but were preceded by a series of observations due to the improvement of telescope technology. In 1781, Herschel discovered a new planet in the solar system that he named George's Star in honour of the king of England, but was subsequently renamed Uranus. Herschel used also his superior telescopes to measure the distance to 100 of stars, using the assumption that all stars emit the same amount of light and that their brightness declines with the square of the distance.

Although he was aware that stars are not equally bright and that his method was inexact, he was confident that he was building an approximate map of the universe. His data implied that the stars were clumped in a disc, which had the form of a flat pancake and was arching around the night sky. This feature of the stars was well known since ancient time and was called by the Romans "Via Lattea" or the Milky Way. Hence the idea of a one island universe was born.

Because the Milky Way supposedly contained all the stars in the universe, the size of the Milky Way was the size of the universe. However, the method employed by Herschel to measure the distances between stars was relative to Sirius, the brightest star in the night sky, which he assumed as a reference. Since he was not able to calculate the distance between Sirius and the Sun, he was not able to calculate the size of the universe.

In 1838, the German astronomer F. W. Bessel became the first person to measure the distance to a star. Bessel used the parallax method. He observed the star Cigny 61 at a certain position of the Earth in her movement around the Sun and he did so along a particular line of sight. He observed the same star 6 months later and noticed that the line of sight has slightly shifted. The right angle triangle form by the Sun, Cigny 61 and Earth allowed him to calculate the distance from the star using simple trigonometry. However, most of the stars in the night sky are so far away that it is not possible to measure the line of sight shifts and hence their distance.

Measuring the distance to very far away stars became possible through the works of Henrietta Leavitt. At the end of 1890s, Leavitt was working for the Harvard College Observatory. She developed a particular curiosity on Cepheid stars. The information available was their period of variation and their brightness. She was aware that one can perceive only the apparent brightness and not the absolute brightness of a star. However, in studying the Cepheid in the Small Magellanic Cloud, she made the assumption that the Cepheid were roughly at the same distance from Earth, since the Small Magellanic Cloud was very far away from earth. Based on this assumption, she realized that the apparent brightness of each Cepheid was a true indication of its absolute brightness in relation to the other Cepheid.

She then plotted a graph of the apparent brightness of 25 Cepheid versus their period of variation and discovered that the longer the period of variation the brighter the Cepheid (Fig. 3.1).

More importantly, she was able to formulate the mathematical equation for this relationship. She also discovered that it was possible to compare any two Cepheid and measure their relative distance from Earth. If one observed two Cepheid with

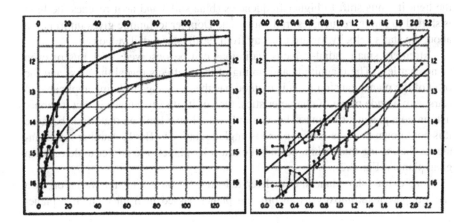

Fig. 3.1 Graph of Leavitt' relationship between brightness and period of variation (http://www. physics.ucla.edu/~cwp/articles/leavitt/leavitt.note.html)

similar period of variation but one being fainter than the other, it meant that it was farther away from Earth. Since brightness fades with the square of the distance, she had the mathematical formula to measure how far away it was from earth.

She lacked however a yardstick. If only the distance of one Cepheid from Earth could be found than it would be possible to anchor Leavitt's measurement scale and estimate the distance for every Cepheid. This was accomplished few years later by Shapley and Herzsprung. The period-luminosity relationship for Cepheid made them the first "standard candle" in astronomy, allowing scientists to compute the distances to stars too remote for stellar parallax observations to be useful. It was now possible to measure the scale of the universe, which was done by Hubble who discovered that it was much bigger that astronomers imagined and that the Milky Way was just one galaxy among numerous others. However the questions of how common were the star systems and how similar were their planets to the ones in the solar system remained unanswered.

3.1 Extra Solar Discoveries

In the second half of the 20th century, thanks to advances in telescope technology both on the ground and in space, astronomers were ready for the discovery of extra solar systems. The first discovery was made at the Observatoire de Haute-Provence[1], which ushered in the modern era of extra solar planet discovery.

Astronomers have detected extra solar planet indirectly by measuring their gravitational influence on the motion of their parent stars (radial velocity method). They used the Doppler Effect to analyze the motion and properties of the star and a planet. Both the planet and the star are orbiting a common centre of mass, meaning that the star and the planet gravitationally attract one another causing the star to wiggle around a point of common centre mass. As the star approach an observer, the light it emits shift to higher frequencies (blue shift) and as it recedes the light is shifted to lower frequencies (red shift). The Doppler technique gives information about the star's velocity toward or away from Earth (star radial velocity).

Extremely small radial-velocity variations can be observed, of 1 m/s or even somewhat less. The magnitude and timing of these effects reveal the mass and the orbit of the planet causing it. This method has the advantage of being applicable to stars with a wide range of characteristics. One of its disadvantages is that it cannot determine a planet's true mass, but can only set a lower limit on that mass, because it is not known how the orbital plane is angled relative to Earth. Another disadvantàge is that, using this technique, small extra solar planets are harder to detect than giant planets (Fig. 3.2).

More extra solar planets were later detected by observing the variation in a star's apparent luminosity as an orbiting planet passed in front of it. If a planet

---

[1] See Mayor and Queloz 1995.

Fig. 3.2 Measuring the gravitational influence (Las Cumbres Observatory Global Telescope Network, Inc.)

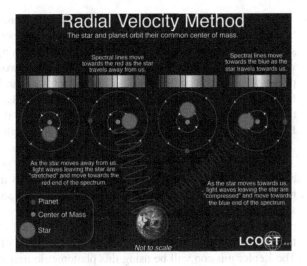

Fig. 3.3 Transit method of detecting extra solar planets (http://upload.wikimedia.org/wikipedia/commons/thumb/8/8a/Planetary_transit.svg/300px-Planetary_transit.svg.png)

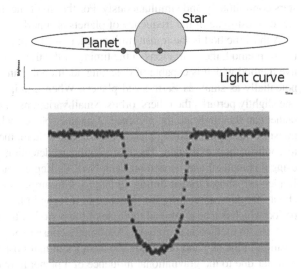

crosses in front of its star's disk, then the observed brightness of the star drops by a small amount. The amount by which the star dims depends on its size and on the size of the planet, among other factors. This has been the second most productive method of detection. The transit method reveals the radius of a planet, and it has the benefit that it sometimes allows a planet's atmosphere to be investigated through spectroscopy (Fig. 3.3).

The transit signal indicates that the bigger are the planets and the smaller are the stars, the higher is the probability of observing them. This method has two major disadvantages.

First, planetary transits are only observable for planets whose orbits happen to be perfectly aligned from the astronomers' vantage point. Less than 1 % of the stars (F-, G and K class dwarfs are the most promising candidate stellar types) would have this kind of desired orbit. If the angle would be different, then the planet would never appear to pass in front of the star. Then, this method would fail. *Second*, the method suffers from a high rate of false detections. Transit detection requires additional confirmation, typically from the radial-velocity method. However, when combined with the radial-velocity method, one can determine the density of the planet, learning something about the planet's physical structure.

One advantage of the transit method is that it makes possible to study the atmosphere of the transiting planet. When the planet transits the star, light from the star passes through the upper atmosphere of the planet. By studying the high-resolution stellar spectrum carefully, one can detect elements present in the planet's atmosphere. A planetary atmosphere could also be detected by measuring the polarization of the starlight as it passed through or is reflected off the planet's atmosphere. The Kepler mission will be using this photometric transit method to find habitable planets around other stars. The Kepler observatory will be studying about 100,000 stars continuously and simultaneously. For the first time astronomers will be able to make an estimate of the frequency of planets around F-, G and K class dwarf stars.

A third method is the transit timing variation. If a planet has been detected by the transit method, then variations in the timing of the transit provide an extremely sensitive method which is capable of detecting additional planets in the system with sizes potentially as small as Earth-sized planets. When multiple planets are present, each one slightly perturbs the others' orbits. Small variations in the times of transit for one planet can thus indicate the presence of another planet, which itself may or may not transit. If multiple transiting planets exist in one system, then this method can be used to confirm their existence. The first significant detection of a non-transiting planet using this method was carried out with NASA's Kepler satellite. The transiting planet Kepler-19b shows transit timing variations with amplitude of 5 min and a period of about 300 days, indicating the presence of a second planet, Kepler-19c, which has a period which is a near rational multiple of the period of the transiting planet.

A fourth method is astrometry. Astrometry consists of precisely measuring a star's position in the sky and observing the changes in that position over time. The motion of a star due to the gravitational influence of a planet may be observable. Astrometry works best when the orbit of the planet around the star is perpendicular to the viewer. If the orbit is edge on, then a shift in the position cannot be measured and this method will be useless. Because the motion is so small, however, this method has not yet been very productive. It has produced only a few disputed detections, though it has been successfully used to investigate the properties of planets found in other ways.

A fifth method is the gravitational microlensing, exploiting an interesting aspect of Einstein's theory of relativity. Gravitational microlensing occurs when the gravitational field of a star acts like a lens, magnifying the light of a distant background star. This effect occurs only when the two stars are almost exactly aligned. Lensing events are brief, lasting for weeks or days, as the two stars and Earth are all moving relative to each other. More than a thousand such events have been observed over the past 10 years.

Fig. 3.4 Confirmed and candidates' extra solar planets. This research has made use of the NASA Exoplanet Archive, which is operated by the California institute of technology, under contract with the National Aeronautics and Space Administration under the Exoplanet Exploration Program (NASA Exoplanet Archive)

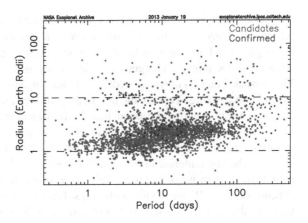

If the foreground lensing star has a planet, then that planet's own gravitational field can make a detectable contribution to the lensing effect. Since that requires a highly improbable alignment, a very large number of distant stars must be continuously monitored in order to detect planetary microlensing contributions at a reasonable rate. This method is most fruitful for planets between Earth and the center of the galaxy, as the galactic center provides a large number of background stars. If astronomers point their telescopes at a region of the sky with lot of stars, every now and then one of them might slowly brighten and dim as another unseen object passes between Earth and the distant star. A planet, orbiting the middle star, causes his slight brightening, which is a blip in the light curve.

As of November 2012, a total of 846 confirmed extra solar planets are listed in the extra solar Planets Encyclopaedia, including a few that were confirmations of controversial claims from the late 1980s. That count includes 665 planetary systems, of which 126 are multiple planetary systems. Most initially confirmed extra solar planets were massive planets that orbited very close to their parent stars. Astronomers were surprised by these "hot Jupiters", since theories of planetary formation indicated that giant planets should only form at large distances from stars.

The goal of discovering Earth-analogs orbiting other stars continues, as the current crop of extra solar planets consists mostly of planets more massive and hotter than Earth as appears clearly in Fig. 3.4.

3.2 Habitable Planets Discoveries

As a rough cut, one may define a "habitable zone"[2] around a star as one where water could exist on the surface of a planet without boiling or freezing. This is an attractive concept because only a planet's orbit needs to be known to establish

---

[2] See Kasting et al. 1993.

whether or not it is habitable. Water's properties are important for life for different reasons. The oxygen atom in the water molecule has a couple of electrons that are not engaged in bonding with hydrogen atom. Their negative charges are however attracted to the positive charges in the hydrogen atoms of other water molecules. These weak bonds, known as hydrogen bonds, makes water the only liquid that expands on freezing. This means that hydrogen does not sink at the bottom of the ocean. If water were as the other liquids, Earth's oceans would be frozen solid. Life would be untenable. Water plays also important roles in biology. Water is essential in processing the information contained in the DNA, and as such it is centre to the phenomenon of life. It carries out this role by allowing organic compounds to react in ways that ultimately allow replication. All known forms of life depend on water. Water is vital both as a solvent in which many of the body's solutes dissolve and as an essential part of many metabolic processes within the body.

In February 2011 the Kepler Space Observatory Mission team released a list of 1,235 extra solar planet candidates, with 54 of those candidates located within the habitable zone.

Six of the candidates in this zone are smaller than twice the size of Earth. These planets have orbits in their parent star's habitable zone, where it should be possible for liquid water to exist and for Earth-like conditions to prevail (Fig. 3.5). Some other planets are giant planets more similar to Jupiter than to Earth; if any of them have large moons, the moons might be a more plausible abode of life.

Over the next decades, space-based constellations of telescopes, like ESA's Darwin Interferometer and NASA's Terrestrial Planet Finder (TPF), will enable astronomers to identify more Earth-like planets orbiting distant stars. The Terrestrial Planet Finder will be equipped with two separate devices: a coronagraph, blocking out the light from the mother star and reducing its light by a factor

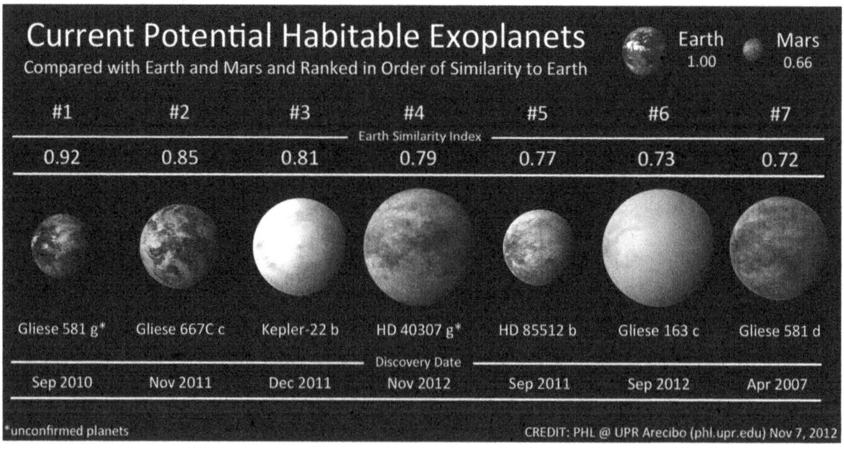

Current Potential Habitable Exoplanets
Compared with Earth and Mars and Ranked in Order of Similarity to Earth   Earth 1.00   Mars 0.66

#1	#2	#3	#4	#5	#6	#7
0.92	0.85	0.81	0.79	0.77	0.73	0.72
Gliese 581 g*	Gliese 667C c	Kepler-22 b	HD 40307 g*	HD 85512 b	Gliese 163 c	Gliese 581 d
Sep 2010	Nov 2011	Dec 2011	Nov 2012	Sep 2011	Sep 2012	Apr 2007

Earth Similarity Index — Discovery Date

*unconfirmed planets   CREDIT: PHL @ UPR Arecibo (phl.upr.edu) Nov 7, 2012

Fig. 3.5 Current potential habitable planets (Planets Habitability Laboratory, University of Puerto Rico, Arecibo)

of a billion; and an interferometer, which uses the interference of light waves to cancel the light from the mother star by a factor of a million.

Although the distance to its star is a significant indicator of the habitability of a planet, other factors are important as well. One important factor is the presence of an atmosphere, containing chemical elements, such as oxygen, which will be necessary to sustain future colonization of the planet. Since oxygen is a very unstable chemical element, its presence in the atmosphere of a planet is a good indicator of the presence of an external agent that maintains the planet's atmosphere in disequilibrium. On Earth, we know that only photosynthesis performed by plants or cynobacteria is maintaining the atmosphere out of equilibrium, but one cannot rule out the presence of other unknown agents in the extra solar planet.

Another important factor is the presence of a moon. On Earth, the presence of the Moon has a powerful stabilizing influence. The Moon gravitational influence allows Earth's obliquity to oscillate by about 1.3° about its mean position of 23.3°. If the Moon did not exist, Earth's obliquity would evolve chaotically, similarly to what happened on Mars. Chaotic evolution of obliquity would have had dire consequence for the climate on Earth and complex life on Earth would not have had time to evolve, hence the importance of the presence of a moon around a planet whose habitability had been determined by its distance to his star.

Finally, the planet should have a magnetic field around it. Absence of a magnetic field would leave the planet at the mercy of fast moving electrically charged particles that are blown towards the planet by its star. This would gradually blow away any atmosphere around the planet, leaving in the end about nothing, as happened to Mars. Although the planet may be habitable in the short term, in the longer term it would become difficult to inhabit by human beings.

In the near future, both terrestrial and space-based telescopes will allow analyzing the spectrum of planets' light to ascertain the presence of other chemical elements, such as oxygen. However, with present technologies, it is not possible to detect the presence of a moon around a planet orbiting its star in the habitable zone, nor can one detect the presence of a magnetic field around it. Only an exploratory probe will be able to gather this information, which will be able to confirm that the planet is suitable for human colonization.

3.2.1 Prime Targets for Interstellar Missions

In our scenario it is important to identify habitable planets within 5/15 light years from Earth. There are many criteria by which to choose the prime targets for an interstellar probe mission. One criterion would be to choose as a target a star similar to our Sun. Since our Sun is positively known to possess a planetary system, it can be assumed that stars similar to the Sun (same spectral type, mass, radius and luminosity) also have some possibility of possessing a planetary system. The argument advanced here is that, by using a star's similarity to the Sun as a selection criterion, we would be choosing target stellar systems not only for their potential planetary systems but also on their ability to support life.

Another criterion is to look at stars different from our Sun. A class of stars, M-class dwarf stars, has attracted the attention of astronomers. These stars are much smaller than the Sun and so thrifty with their fuel resources that can last trillion of years. They are more numerous than heavier Sun-like stars. Hence these stars could be ideal in the search of habitable planets. However, they present a problem, which is still debated by astronomers. Since their habitable zone is 5 times narrower than the one around the Sun, it is uncertain that planets within this zone could maintain stable orbits over long periods of time. If a planet orbit is too closely to a star, it could become tidally locked after few billion years from its formation, thus presenting the same face inward that could make it not hospitable for life. However, some astrobiologists maintain that it is possible for liquid water to exist in the twilight zone between the face permanently exposed to its sun and the dark face. The verdict on this issue is still open.

At the moment, there are seven prime targets within 15 light years of the Sun (Table 3.1). This list contains three G-class stars and several M-class dwarfs. As we learn more about the formation of planetary systems and we search the neighbouring stellar systems for planets, this list will change. The main characteristics of these stellar systems are summarized in Table 3.1. Astronomical observations should be targeted to these star systems to confirm: (1) the presence of planets orbiting the stars in the habitable zone; (2) whether or not these star systems have multiple planets; and (3) the presence in their atmosphere of chemical elements necessary to support life.

Estimate of the age of the Alpha Centauri system shows that it is likely to be slightly older than the Solar System, with values ranging from 4.5 to 7

Table 3.1 Prime targets for interstellar missions

Stellar system	Distance in light years from Earth	Remarks
Alpha Centauri	4.3	Closest system. Triple (G0, K5, M5). Component almost identical to Sun. High probability of "life bearing" planets
Barnard's Star	6	Closest system known to have one, and perhaps two or more planetary companions. Very small, low luminosity red dwarf (M5)
Lalande 21185	8.2	Red dwarf star (M2) known to have a planet
Epsilon Eridani	10.8	Single star system; slightly smaller and cooler than the Sun (K2), may have a planetary system similar to the solar system
Procyon	11.3	Large, hot white star (F5), second only to Altair in luminosity (within 20 light years). System contains small white dwarf
Tau Ceti	11.8	Single star system, similar in size and luminosity to the Sun (G4). High probability of possessing a "Solar-like" planetary system
Gliese 876	15	Red dwarf star with confirmed planetary system composed of four planets

Fig. 3.6   Alpha Centauri
(http://earthsky.org)

billion years. According to widely accepted models of planetary formation, Alpha
Centauri planets should be nearly as old. The orbit of the planets is not believed to
be destabilized by the influence of companion star Alpha Centauri A, which has a
periastron (closest approach) distance a bit larger than the radius of Saturn's orbit.
Scientists say the Alpha Centauri system, particularly Alpha Centauri B, has the
ingredients for an Earth-like planet. Much of its matter is made up of elements
heavier than hydrogen and helium, so there would be plenty of heavier material to
make planets from.

As of October 2012, a planet Alpha Centauri Bb has been discovered. It is
Earth-size planet, but it does not lie in the habitable zone within which liquid
water can exist. The astronomers who found it say that it is likely there are other
planets circling the same star, a little farther away where it may be cool enough
for liquid water. These hypothetical companions are likely to have wider orbits,
and would be difficult to find with current instruments. Detecting additional planets in the system will become easier when the ESO's next-generation spectrometer
ESPRESSO comes online in 2017. ESPRESSO is specifically designed to look for
Earth-like planets, and will provide radial-velocity measurements several times
more precise than those used to find Alpha Centauri Bb (Fig. 3.6).

Barnard's Star is a very low-mass red dwarf star about six light-years away
from Earth. At seven to 12 billion years of age, Barnard's Star is considerably
older than the Sun and it might be among the oldest stars in the Milky Way galaxy. For a decade from 1969 to about 1973, a substantial number of astronomers
accepted a claim by Peter van de Kamp (1969) that detected, by using astrometry,
a perturbation in the proper motion of Barnard's Star consistent with its having
one or more planets comparable in mass with Jupiter. Other astronomers subsequently repeated Van de Kamp's measurements, and two important papers in 1973
undermined the claim of a planet or planets.

Null results for planetary companions continued throughout the 1980s and
1990s, the latest based on interferometer work with the Hubble Space Telescope

in 1999. While this research has greatly restricted the possible existence of planets around Barnard's Star, it has not ruled them out completely. NASA's Terrestrial Planet Finder and ESA's similar Darwin interferometry mission could accomplish these detections.

Lalande 21185 is a red dwarf star in the constellation of Ursa Major. Although relatively close by, it is only magnitude 7 in visible light and thus is too dim to see with the unaided eye. At approximately 8.31 light-years away from Earth, this star is the fourth closest stellar system to the Sun; only the Alpha Centauri system, Barnard's Star and Wolf 359 are known to be closer. In 1996, George Gatewood announced the discovery of multiple planets in this system, detected by astrometry. He claimed that such planets would usually appear more than 0.8 arcs second from the M dwarf itself. However subsequent searches by others, using coronagraphs and multifilter techniques to reduce the scattered-light problems from the star, have yet to positively identify any such planets.

Epsilon Eridani is a star in the southern constellation Eridanus. At a distance of 10.5 light years from Earth, it has an apparent magnitude of 3.73. It is the third closest of the individual stars to the Sun. It was the closest star known to host a planet until the discovery of Alpha Centauri Bb. Its age is estimated at less than a billion years. Because of its youth, Epsilon Eridani has a higher level of magnetic activity than the present-day Sun, with a stellar wind 30 times as strong (Fig. 3.7).

Epsilon Eridani is smaller and less massive than the Sun, and has a comparatively lower level of elements heavier than helium. Astronomers categorize it as a main-sequence star of spectral class K2, which means that energy generated at the core through nuclear fusion of hydrogen is emitted from the surface at a temperature of about 5,000 K, giving the star an orange hue.

As one of the nearest Sun-like stars, Epsilon Eridani has been the target of many attempts to search for planets. However, its chromospheric activity and variability means that finding planets with the radial velocity method is difficult,

Fig. 3.7 Epsilon Eridani planets (Astrophysical Institute and University Observatory, Friedrich-Schiller-Universität Jena)

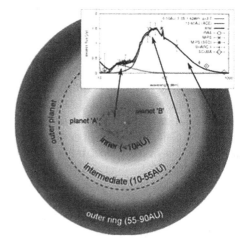

Fig. 3.8 Tau Ceti system
(Media INAF)

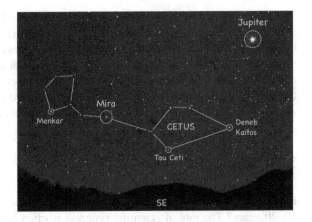

because the stellar activity may create signals that mimic the presence of planets. Attempts at direct imaging of potential extra solar planets have proven unsuccessful to date. Infrared observation has shown there are no bodies of three or more Jupiter masses in this system.

Tau Ceti is a star in the constellation Cetus that is spectrally similar to the Sun, although it has only about 78 % of the Sun's mass. At a distance of just under 12 light-years from the Solar System, it is a relatively nearby star. The star appears stable, with little stellar variation. Astrometric or radial velocity measurements have not yet detected planets around Tau Ceti. Because of its debris disk, any planet orbiting Tau Ceti would face far more impact events than the Earth.

Principal factors driving research interest in Tau Ceti are its Sun-like characteristics and their implications for possible planets and life. Tau Ceti was a target of a few radial velocity planetary searches, which have failed to find any periodical variations attributable to a planet. The velocity precision reached so far is about 11 m/s measured over a 5 year time span. This result excludes the presence of hot Jupiters, but Earth-like planet in orbit around the star is not precluded. If "hot Jupiters" did exist in close orbit, they would likely disrupt the star's habitable zone; their exclusion is thus a positive indication of the possibility of Earth-like planets. Recently, a team of astronomers[3] have identified five planets orbiting Tau Ceti. According to this study, these planets have masses from two to six time the Earth's mass. The smallest of these planets is located in the habitable zone (Fig. 3.8).

Gliese 876 is a red dwarf star approximately 15 light-years away from Earth in the constellation of Aquarius. It is the fourth closest known star to the Sun confirmed to possess a planetary system (after Alpha Centauri, Gliese 674 and Epsilon Eridani) and the closest such system known to consist of multiple planets. Gliese 876 d is the third planet discovered orbiting the red dwarf star Gliese 876.

---

[3] See Butler P. et al. 2012.

At the time of its discovery, the planet had the lowest mass of any known extra solar planet. Due to this low mass, it can be categorized as a super-Earth. Since Gliese 876 d has only been detected indirectly by its gravitational effects on its star, properties such as its radius, composition, and temperature are unknown. This type of massive terrestrial planet could be formed in the inner part of the Gliese 876 system from material pushed towards the star by the inward migration of the gas giants. Astronomers do not rule out the presence of other planets in the habitable zone.

3.3 Conclusions

How likely is that we will discover a habitable planet in the course of the next 10–20 years? The rate of scientific progress is often hard to measure. But in certain circumstances, the data are unambiguous and easy to measure, creating a trend. And when that happens, it is possible extrapolating and predicting the way things will be. The discoveries of extra solar planets follow a well understood pattern. The first extra solar planets were massive, many times the size of Jupiter, and so easier to spot. As techniques have improved, however, astronomers have found smaller planets, some just a few times more massive than Earth. At present, astronomers have detected planets orbiting their suns in the habitable zone, where liquid water may be present. The presence of liquid water depends on the size of the star, and the conditions on the surface, such as the amount of greenhouse effect. In the near future, advancement in observation techniques will allow whether these conditions are present in the planets situated in the habitable zone.

Two scientists[4] have recently written a paper on this subject. They base their projection on math and trends from the past 15 years of extra solar discoveries. In their paper they indicate that have taken this data and projected it forward to see when an Earth-like planet is likely to crop up. The results have a heavy-tailed distribution in which there is a 66 % probability of finding the other Earth by 2013, a 75 % probability by 2020 but a 95 % probability by 2264. Although these forecasts are to be taken cautiously, there is no real dispute among astronomers that the discovery of an Earth-like planet is on the cards.

Box 3.1
Finding Masses of Extrasolar Planets

The Doppler technique gives us information about the star's velocity toward or away from Earth, and from this we can find its mass. Consider a system containing one planet and one star. One often says that the planet is orbiting the star; however, it is really orbiting the centre of mass for the planet-star system.

---

[4] See Abersman and Laughlin 2010.

If these objects have the same orbital period, then the centre of mass does not move and the system does not experience a change in momentum; therefore, the planet and the star must have equal magnitudes of momentum, and can be set equal to each other. Momentum is mass time velocity:

$$M_{star}v_{star} = M_{planet} * v_{planet}$$

where M is the mass of the star or planet, and v is the velocity relative to the centre of mass of the system. Solving for the mass of the planet:

$$M_{planet} = (M_{star} * v_{star})/v_{planet}.$$

We know the mass and velocity of the star from the Doppler technique. We also know that distance is equal to rate x time, and rate = distance/time. For simplicity, even though orbits are generally elliptical, we can use the circumference of a circle, $2\pi r$, for the distance travelled by a planet around its star. In this case, we will substitute a, the semi major axis of the orbit, for r. We use the planet's orbital period, p, for the time it takes the planet to travel around its star. We know the orbital period, p, from the Doppler technique. Thus, the planet's average orbital velocity is its average distance travelled divided by the time it takes to travel around the star.

$$v_{planet} = 2\pi a_{planet}/p_{planet}.$$

Substituting this equation above, we find that the planet's mass is given by:

$$M_{planet} = M_{star}v_{star} p_{planet}/\pi a_{planet}.$$

You can now see how easy it is to find the mass of a planet, just by taking a few spectral measurements of the star it orbits.

References

Abersman S., Laughlin G.: A scientometric prediction of the Discovery of the First potentially habitable planet with a mass similar to Earth. arXiv: 1009.2212 [astro-ph.EP] (2010)

Butler P. et al.: Signals embedded in the radial velocity noise. Periodic variations in the τ Ceti velocities. Astronomy and Astrophysics, Dec (2012)

Kasting, J.F., et al.: Habitable zones around main sequence stars. Icarus **101**, 108–128 (1993)

Leavitt H.S.: Periods of 25 Variable Stars in the Small Magellanic Cloud. Harvard College Observatory Circular, pp. 173 (1912)

Mayor, M., Queloz, D.: A Jupiter-mass companion to a solar-type star. Nature **378**(6555), 355–359 (1995)

Van de Kamp, P.: Alternate dynamical analysis of Barnard's star. Astron. J. **74**(8), 757 (1969)

Chapter 4
Interstellar Exploration Technologies

I have learned to use the word 'impossible' with the greatest caution.

Wernher von Braun

Astronautics and astronautical engineering are the theory and practice of navigation beyond the Earth's atmosphere. The early history of astronautics is theoretical. The fundamental mathematics of space travel was established by Newton in his 1687 treatise "Philosophiae Naturalis Principia Mathematica". Other mathematicians, such as Leonhard Euler and Joseph Louis Lagrange also made essential contributions in the 18th and 19th centuries.

Astronautic developed rapidly in the 20th century because of the theoretical work of Konstantin Tsiolkovsky, and of the early experimentations of R. H. Goddard in the United States and W. Von Braun in Germany. Konstantin Tsiolkovsky was a rocket scientist and a pioneer of the astronautic theory. Starting in 1896, Tsiolkovsky had systematically studied the theory of motion of jet apparatus. Thoughts on the use of the rocket principle in the cosmos were expressed by him as early as 1883, and a rigorous theory of jet propulsion was developed in 1896. Tsiolkovsky derived the formula, called the "formula of aviation" by him, establishing the relationship between: speed of a rocket at any moment; specific impulse fuel; and mass of the rocket in the initial and final time. Tsiolkovsky stated that he developed the theory of rocketry only as a supplement to philosophical research on the subject. Tsiolkovsky championed the idea of the diversity of life in the universe and was the first theorist and advocate of human space exploration.

R. H. Goddard was an American physicist and inventor, credited with creating and building the world's first liquid-fuelled rocket. Goddard's work as both theorist and engineer anticipated many of the developments that were to make spaceflight possible. Two of Goddard's 214 patents—one for a multi-stage rocket design (1914), and another for a liquid-fuel rocket design (1914)—are regarded as important milestones toward spaceflight. Goddard successfully applied three-axis control, gyroscopes and steerable thrust to rockets, all of which allow them to be controlled effectively in flight. He successfully launched it on March 16, 1926. Goddard and his team launched 34 rockets between 1926 and 1941. Goddard received little public support for his research during his lifetime. Although his

G. F. Bignami and A. Sommariva, *A Scenario for Interstellar Exploration and Its Financing*, SpringerBriefs in Space Development, DOI: 10.1007/978-88-470-5337-3_4,

work in the field was revolutionary, he was sometimes ridiculed in the press for his theories concerning spaceflight. Years after his death, he came to be recognized as one of the founding fathers of modern rocketry. One of the most important NASA Research Canters (located in Greenbelt, MD) is now named after him.

Wernehr von Braun was a German-American rocket scientist, aerospace engineer, and space architect. Wernher von Braun was born in Wirsitz in 1912 and was the second of three sons. He belonged to an aristocratic family, inheriting the German title of Freiherr, the equivalent to Baron. In 1929, he acquired a copy of Die Rakete zu den Planetenräumen (Rocket into Interplanetary Space) by rocket pioneer Hermann Oberth[1]. Space travel had always fascinated von Braun, and from then on he applied himself to physics and mathematics to pursue his interest in rocket engineering. In 1930 he attended the Berlin Institute of Technology, where he joined the Verein für Raumschiffahrt (VfR), the "Spaceflight Society" and assisted Willy Ley in his liquid-fuelled rocket motor tests in conjunction with Hermann Oberth.

Von Braun had an ambivalent and complex relationship with the regime of the Third Reich. He officially applied for membership in the Nazi Party on November 12, 1937. He became one of the leading figures in the development of rocket technology in Nazi Germany during World War II. In the late 1930s, von Braun was the central figure in Germany's rocket development program, responsible for the design and realization of the V-2 combat rocket during World War II. As for his attitude toward the National Socialist regime in the late 1930s and early 1940s, there can be little doubt that he was a loyal, perhaps mildly enthusiastic subject of Hitler's dictatorship.

In the spring of 1945, von Braun assembled his planning staff and asked them to decide how and to whom they should surrender. Afraid of the well known Soviet cruelty to prisoners of war, von Braun and his staff decided to try to surrender to the Americans. The American high command was well aware of how important their catch was: von Braun had been at the top of the Black List, the code name for the list of German scientists and engineers targeted for immediate interrogation by U.S. military experts. On June 20, 1945, the U.S. Secretary of State approved the transfer of von Braun and his specialists to America, where they worked initially for the United States Army.

The early history of astronautics was dominated by the development of propulsion systems based on chemical fuels, both liquid and solid, and by its strict relationships to military programs. The rocket development program of Germany in the 1930s and during the Second World War was centred in the development of the V2 combat rocket. When von Braun surrendered to the Americans at the end of the Second World War, he was initially employed the U.S. Army for the development of the Jupiter rocket. Between 1950 and 1956, von Braun led the Army's rocket development team at Redstone Arsenal, resulting in the Redstone rocket, which was used for the first live nuclear ballistic missile tests conducted by the United

---

[1] See Oberth 1964.

States. This led to development of the first high-precision inertial guidance system on the Redstone rocket.

At the time of the launch of the Sputnik by the Soviet Union in 1957, the military establishment in the United States was determined to the lead the space race. However, the Army, the Navy and the Air Force jumped into the race without any direction. Eisenhower could see the potential internal arms race. He proposed to outlaw the very concept of space weapons. He said that "outer space should be used only for peace purposes". In 1958, he called for the creation of a National Aeronautics and Space Administration (NASA). All space programs were to be turned over to NASA, including the Army's Jet Propulsion Laboratory at Cal Tech.

It was only after President Eisenhower decided to concentrate all space programs in the civilian NASA that civilian space programs started. Von Braun became a central figure in these development programs, which culminated in the Saturn rocket and the Apollo program for landing men on the Moon. The Saturn family of American rocket boosters was developed to launch heavy payloads to Earth orbit and beyond. The challenge that President John F. Kennedy put to NASA in May 1961 to put an astronaut on the Moon by the end of the decade put a sudden new urgency on the Saturn program.

Originally proposed as a military satellite launcher, the Saturn family of rockets were adopted as the launch vehicles for the Apollo moon program. The two most important members of the family were the Saturn IB and the Saturn V. Saturn IB is a refined version of the Saturn I with a more powerful first stage. These carried the first Apollo flight crew, plus three Skylab and one Apollo-Soyuz crews, into Earth orbit. The Saturn V is the only launch vehicle to transport human beings beyond Low Earth Orbit. A total of 12 humans set foot on the Moon in the four-years spanning July 1969 through December 1972.

Von Braun had more ambitious programs in mind. In 1969, he presented to the US Congress a plan to land man on Mars. This plan was based on his early visionary publication: *The Mars Project*. The plan consisted in the building of a space station, where the star ships would have been assembled, and the launching of ten star ships with a total crew of seventy persons. Although von Braun spoke of nuclear propulsion in his version of the Mars Project, his proposal was finally based on the well known chemical propulsion system. However, this proposal did not pass the approval of Congress for a few votes. Although NASA and afterward ESA continued to dedicate some resources to the development of new propulsion systems, the lack of political will brought effectively to a halt of manned exploration of space.

4.1 Propulsion Theory

For purposes of this chapter, we will concentrate on Konstantin Tsiolkovsky's equation. This equation makes it possible to compute the final velocity of a spaceship from its mass, the combined mass of propellant and spaceship, and exhaust velocity of the propellant. The first two factors depend strongly on the details of

Table 4.1 Specific impulses in seconds

Type of spaceship engine	Specific impulse
Solid fuel	250
Liquid fuel	450
Ion	3,000
Nuclear fission	800–1,000
VASIMIR plasma	1,000–30,000
Nuclear fusion	2,500–200,000
Nuclear pulse	10,000–1 million
Matter-antimatter annihilation	1 million–10 million

engineering and construction. When considering space propulsion for the future, it seems appropriate to defer the study of such specifics. Thus, exhaust speed is seen as the main focus of long-term advances in propulsion technology.

Since interstellar distances are very great, a tremendous velocity is needed to get a spaceship to its destination in a reasonable amount of time. Acquiring such a velocity on launch and getting rid of it on arrival will be a formidable challenge for spaceship designers. We assume that spaceships for interstellar exploration will be assembled on a space-port in orbit around the Earth or in the Lagrangian librations point between Earth and the Moon. Therefore, the problem of escaping Earth's gravity is bypassed. We have to consider propulsion systems capable of accelerating the spaceship in space.

This chapter will explore whether the propulsion technology for travelling at a significant fraction of the speed of light[2] could be available in the foreseeable future, let say within the next 50/100 years. The equations describing the optimum design of a propulsion system required to reach near relativistic velocities can be very complicated (see Box 4.1). However, the basic features of the desired propulsion system can be explained without equations. When in space, the purpose of a propulsion system is to change the velocity of a spaceship. In order for a spaceship to work, two things are needed: reaction mass and energy. In space, the spaceship must bring along some mass to accelerate away in order to push itself forward in accordance with Newton's third law: "For every action there is an equal and opposite re-action".

Such mass is called reaction mass. Reaction mass must be carried along with the spaceship and is irretrievably consumed when used.

When discussing the efficiency of a propulsion system, designers often focus on effectively using the reaction mass. One way of measuring the amount of impulse that can be obtained from a fixed amount of reaction mass is the specific impulse, i.e. thrust deliverable ejecting a unit of weight of the propellant. Usually it is measured as the impulse per unit weight-on-Earth. The unit for this value is seconds. Table 4.1 shows the specific impulses, measured in seconds, for different kinds of proposed spaceship engines. It shows that the most efficient engines are the ones based on nuclear fusion, nuclear pulse, and matter-antimatter

---

[2] See Norem 1969.

annihilation. Laser sails and ram-jet engines have infinite specific impulse because they do not contain propellant at all.

Since the weight on Earth of the reaction mass is often unimportant when discussing vehicles in space, specific impulse can also be discussed in terms of impulse per unit mass. This alternate form of specific impulse uses the same units as velocity (m/s), and in fact it is equal to the effective exhaust velocity of the engine. One might think that, if the spaceship effective exhausts velocity were a fraction of the speed of light, the fastest the spaceship could move would be that same fraction of the speed of light.

However, a spaceship is propelled forward by a thrust force equal in magnitude, but opposite in direction, to the mass flow rate of the propellant multiplied by the speed of the exhaust particles relative to the spaceship. Thus, a spaceship can attain a final velocity that is higher than the velocity of the particles in its exhaust jet. However, the energy required to generate that impulse is proportional to the exhaust velocity, so that more mass-efficient engines require more energy, and are typically less energy efficient. This is a problem if the engine is to provide a large amount of thrust. To generate a large amount of thrust, it must use a large amount of energy per second. Therefore high-mass-efficient engines require large amounts of energy per second to produce high thrusts.

This is not the case if the propulsion system allows choosing the propellant mass and energy source separately, as for present day solar electric, the proposed laser heated rockets, and matter-antimatter annihilation propulsion systems. For these types of propulsion systems, calculations show that the optimum exhaust velocity is about 6/10 of the final velocity of the spaceship and the mass ratio (the ratio of a spaceship's *wet mass*—vehicle plus payload plus propellant—to its *dry mass*—vehicle plus payload) is about 5. That is, for every kilogram of payload, you need only four kilograms of propellant, or a total mass ratio of 5 kg. If you want to decelerate at the target star, then this requires a mass ratio of 5 in each stage, or a total launch mass ratio of 25. A mass ratio of 25 is not a high mass ratio, and most of present space shots have used much higher ones. The Saturn V mass ration was about 1,000.

In other propulsion systems, where one is not allowed choosing the exhaust velocity arbitrarily and where the source of exhaust particles and the source of energy are one and the same, as in present day chemical rockets and the proposed fusion and nuclear pulse propulsion systems, the exhaust velocity is determined by the characteristics of the fuel. Typically, these exhaust velocities are lower than one would like them to be in order to attain high trust, and therefore these spaceships require very large mass ratios.

4.2 Propulsion Systems

Several new and promising concepts are being investigated providing fast space transportation. NASA, ESA and other space agencies have been engaging in research into these topics for several years. At present, some of the best technology candidates for near relativistic space engine include: nuclear power, anti-matter systems, and laser driven sail starships.

Other technologies allowing travelling at velocities "higher" than the speed of light have been explored at the theoretical level. In general relativity, space-time becomes a "fabric" and this fabric can stretch faster than the speed of light. In this case, the spaceship will not move at all, but will be transported by the stretching of space in a twinkling of an eye to enormous distances. It can also allow for holes in space through which one can take short-cuts though space and time. Stretching space was analyzed by M. Alcubierre's warp drive.[3] The key to Alcubierre drive is the energy necessary to stretch space-time, which would allow the spaceship to cover enormous distances instantaneously. According to his calculations, one would need the presence of dark energy, one of the most exotic entities in the universe.

Space ripping was first introduced by Einstein in 1935 through the concept of wormholes. The wormhole is a device that can connect to distant points in space. In classical geometry, the shortest distance between two points is a straight line. However, in curved space, the shortest distance between two points is a wormhole. In 1994, K. Thorne and his colleagues at Cal Tech examined a theoretical example of a traversable wormhole, that is one in which a spaceship could pass freely back and forth. However, in order to keep the opening of the wormhole stable, dark energy is needed. Thus, both concepts require the presence of dark energy.

Although the existence of dark energy is implied by the astronomical observations of the accelerating expansion of the universe, physicists have no idea of what dark energy is. Since we do not know the physics of dark energy, it is impossible to say when these technologies could be realized. Scientific breakthroughs are unpredictable. However, even if a breakthrough in our understanding of dark energy is assumed, realization of these technologies will require an enormous amount of energy. Harnessing energy at this scale makes these technologies possibly feasible only in the very distant future, well outside the horizon of this scenario. Therefore, we have excluded them from it.

4.2.1 Nuclear Based Engines

The fundamentals, problems and potentialities of spaceship propulsion systems powered by nuclear reactors have been extensively discussed in the technical literature.[4] Both fission and fusion appear promising for space propulsion applications, generating higher velocities with less reaction mass than the current generation of chemical rockets.

---

[3] See Alcubierre 1994.

[4] See Bussard and Delauer 1958; and Bussard 1960.

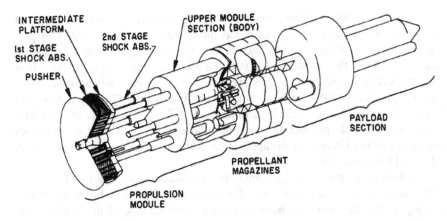

Fig. 4.1   Project orion configuration (http://upload.wikimedia.org/wikipedia/commons/thumb/7/76/ProjectOrionConfiguration.png/480px-ProjectOrionConfiguration.png)

4.2.1.1 Nuclear Fission

Nuclear fission thermal engines are most advanced in their design and are the most powerful rocket motors devised to date. An early example of nuclear fission propulsion system was developed in the early 1960's at Gulf General Atomic (Project Orion). The propulsion system operates simply by jettisoning a nuclear bomb, exploding it, and absorbing part of the momentum of the resulting debris. Freeman J. Dyson[5] has reported a generalized pulse system which will give the reader some idea of its potentials and drawbacks.

Dyson's nuclear pulse vehicle (Fig. 4.1) is able to maintain one g acceleration for about 10 days, when its bomb supply will be exhausted. At this time, it would have reached a velocity of 10,000 km/s (3.3 % of the speed of light). This would allow exploration of the solar system, such a mission to Mars, but the velocity of the spaceship is not adequate for interstellar exploration in a reasonable amount of time. The development of this propulsion concept was practically halted at the time of the signing of the Nuclear Weapons Test Ban Treaty, prohibiting the explosion of nuclear weapons in space.

Variants of the nuclear pulse technique such as those considered for Project Daedal us, using small pellets of deuterium ignited to fusion temperatures by pulsed lasers or by relativistic electron beams have been subsequently developed. Daedalus would be assembled in Earth orbit and have an initial mass of 54,000 t, including 50,000 t of fuel and 500 t of scientific payload. Daedalus was to be a two-stage spaceship. The first stage would operate for 2 years, taking the spaceship to 7.1 % of light speed, and then after it was jettisoned the second stage would fire for 1.8 years, bringing the spaceship up to about 12 % of light speed.

---

[5] See Freeman 1968.

An interesting proposal for a nuclear fission propulsion system was carries out by Rubbia and his colleagues, and it was incorporated in Project 242. This propulsion system is based on the principle that direct energy transfer from fission fragments (FF) to the propulsive gas is by far more efficient than using basic heat exchanges. Many potentially interesting fissionable fuel elements have been surveyed, trying to identify the best choice according to the following criteria: (1) large fission cross section; (2) high fission probability; (3) acceptably long lifetime; (4) reasonable production procedure. On this basis, at least two isotopes appeared to be of interest for the application, namely in order of merit Americium 242 and Plutonium 239.

The heating of the gas is performed by the ionisation energy loss of the FF, which are emitted by a fissile layer. The fissionable layer will be arranged as a coating on the inner walls of chambers trough which the propellant gas flows. The specific ionisation losses in the gas are more than a factor 20 larger than in the fuel layer. Thus, a modest thickness of gas is sufficient to extract most of the energy from the FF. As a consequence, even with very low gas pressure (few bars), the FF can be completely absorbed in a range of few tens of centimetres.

A simple gas dynamic expansion through a nozzle delivers remarkable performances if compared to existing propulsion motors. This essentially because a few kilograms of nuclear fuel are able to generate an amount of energy that is many times the one produced by the largest existing chemical rockets. One advantage of this propulsion system is that it allows the saving of large amount of fuel with respect to other nuclear fission propulsion systems. Bignami et al. have calculated that a round trip to Mars would involve only a few kilos of Americium or Plutonium.

Based on preliminary studies and on a first optimisation for the reference mission described later, a proposed motor configuration has been defined. It consists of a certain number of tubes packed together. Each tube is coated with a thin fissionable layer and opens at one end to let the gas to escape. The tubes are located inside a neutron reflecting structure, which ensures neutron containment and criticality. Some control rods are necessary to control the neutron multiplication. They are inserted in the reflector.

A cooling system to keep the americium foil and the neutron diffuser to the operating temperature could flow around the tubes and transfer the produced heat to a radiating panel. There are numerous engineering problems yet to be solved. The main one is that we do not yet know how to construct tubes and nozzles capable of resisting to temperatures up to 10,000°. Furthermore, the problem of cooling is not yet solved, although in space this may be a solvable problem, since the temperatures in space are very low (2.7° above absolute zero) (Fig. 4.2).

Further improvement of the nuclear fission propulsion was carried out by the Idaho National Engineering Laboratory and Lawrence Livermore National Laboratory with the design of a fission-fragment engine. The fission-fragment engine is an engine design[6] that directly harnesses hot nuclear fission products for thrust, as

---

[6] See Clark and Sheldon 2005.

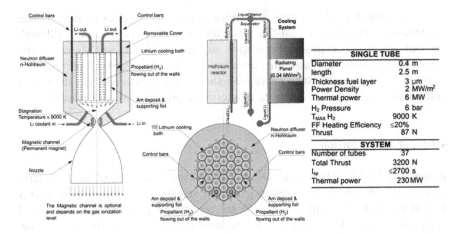

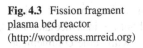

SINGLE TUBE	
Diameter	0.4 m
length	2.5 m
Thickness fuel layer	3 μm
Power Density	2 MW/m²
Thermal power	6 MW
H_2 Pressure	6 bar
T_{MAX} H_2	9000 K
FF Heating Efficiency	≤20%
Thrust	87 N
SYSTEM	
Number of tubes	37
Total Thrust	3200 N
I_{sp}	≤2700 s
Thermal power	230 MW

Fig. 4.2 Layout of the Rubbia motor configuration (Bignami et al. 2011)

Fig. 4.3 Fission fragment
plasma bed reactor
(http://wordpress.mrreid.org)

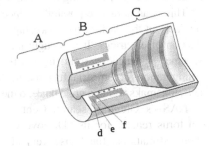

opposed to using a separate fluid as working mass. In a conventional nuclear reactor this high kinetic energy of the fission fragments is dissipated by collisions to generate heat, which is converted to electrical power with efficiencies of no more than 50 %. Alternatively, the fission fragments produced in the plasma reactor can be used directly for providing thrust. This design can, in theory, produce very high specific impulses, while still being well within the abilities of current technologies. In Fig. 4.3, **A** are the fission fragments ejected for propulsion, **B** is the reactor, **C** are the fission fragments decelerated for power generation, **d** is the moderator, either beryllium or lithium, **e** is the containment field generator, and **f** is the induction coil.

In the design of these engines, the fuel is placed into a number of very thin carbon bundles, each one normally sub-critical. Bundles were collected and arranged like spokes on a wheel, and several such wheels were stacked on a common shaft to produce a single large cylinder. The entire cylinder was rotated so that some bundles were always in a reactor core where additional surrounding fuel made the bundles go critical. The fission fragments at the surface of the bundles would break free and be channelled for thrust, while the lower-temperature un-reacted fuel would eventually rotate out of the core to cool. The system thus automatically "selects" only the most energetic fuel to become the working mass.

The efficiency of the system is surprising; specific impulses of greater than 100,000 s are possible using existing materials. This is high performance, although not that which the antimatter propulsion engines could achieve. Moreover, the weight of the reactor core and other elements would make the overall performance of the fission-fragment system lower. Nonetheless, the system provides the sort of performance levels that would make an interstellar precursor mission possible.

4.2.1.2 Nuclear Fusion

A fusion rocket[7] is a theoretical design that could provide efficient and long-term acceleration in space. The present extensive scientific and engineering efforts around the world on controlled fusion are concentrating on the development of "magnetic bottles" to contain high temperature fusion plasma, trying to solve the problem of making the magnetic bottle leak proof. However, for the propulsion application, what is needed is a "leaky" magnetic bottle, the hot plasma emitted from the "leaky" providing the desired rocket thrust.

This is an attractive possibility since it would simply direct the exhaust of fusion product out the back of the spaceship to provide thrust without the intermediate production of electricity. It is thought that this would be easier with some confinement schemes (magnetic mirrors) than with others (tokamaks). The most widely studied configuration for terrestrial fusion is the tokamak. Currently, owing to their bulky magnets, tokamaks weigh a great deal, so the thrust to weight ratio would seem a problem.

NASA's Glenn Research Centre has proposed a small aspect ratio spherical torus reactor for its "Discovery II" conceptual vehicle design. According to their estimations, the "Discovery II" could deliver a manned 172,000 kg payload to Jupiter in 118 days, using 861 metric tons of hydrogen propellant, plus 11 metric tons of Helium-3-Deuterium fusion fuel. The hydrogen is heated by the fusion plasma debris to increase thrust. The exhaust velocity is estimated at 348/463 km/s, or 1/1,000 of the speed of light.

The main alternative to magnetic confinement is inertial confinement fusion (ICF), such as that proposed by Project Daedalus. A small pellet of fusion fuel would be ignited by an electron beam or a laser. To produce direct thrust, a magnetic field would form the pusher plate. In the 1980s, Lawrence Livermore National Laboratory and NASA studied an ICF powered "Vehicle for Interplanetary Transport Applications" (VISTA). The conical VISTA spaceship could deliver a 100 t payload to Mars orbit and return to Earth in 130 days or to Jupiter orbit and back in 403 days. 41 t of deuterium/tritium fusion fuel would be required, plus 4,124 t of hydrogen expellant. The exhaust velocity would be 157 km/s.

A relatively new approach, the magnetic target fusion (MTF), combines the best features of the more widely studied magnetic confinement fusion and inertial confinement fusion approaches. Like the magnetic approach, the fusion fuel is

---

[7] See Matloff and Chiu 1970.

confined at low density by magnetic fields while it is heated into plasma, but like the inertial confinement approach, fusion is initiated by rapidly squeezing the target to dramatically increase fuel density, and thus temperature. MTF uses "plasma guns" (i.e. electromagnetic acceleration techniques) instead of powerful lasers, leading to low cost and low weight compact reactors. The NASA Human Outer Planets Exploration (HOPE) group has investigated a manned MTF propulsion spacecraft. This design would thus be considerably smaller and more fuel efficient due to its higher exhaust velocity (Isp = 700 km/s) than the previously mentioned "Discovery II", "VISTA" concepts (Fig. 4.4).

It is difficult to predict at present whether one or another of the propulsion systems described above are engineering feasible. It is not yet clear whether the inertial confinement fusion is technically feasible, casting doubts on the feasibility of the magnetic target fusion, which has the highest exhaust velocity. Another problem to be overcome is the efficient utilization of the several different forms of energy released during a fusion reaction. Only about 20 % of the released energy will be contained in the kinetic energy of the fusion particles. Ten per cent is in the form of IR-UV radiation, but 70 % is released as X-rays. Some authors have analyzed how to reclaim this X-ray energy by using auxiliary laser thrusters, powered by the waste X-rays. If the efficiency of the laser's X-ray to collimated light conversion process is around 40 %, the laser aided fusion rocket can achieve a velocity of about 10 % of the speed of light with a mass ratio of 20.

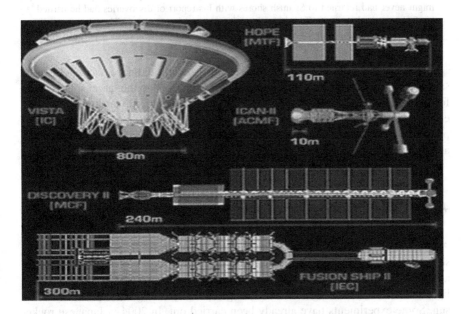

Fig. 4.4 Montage of fusion-powered interplanetary spaceships concepts from 1987 to 2004. *Sources* VISTA (Lawrence Livermore National Laboratories, 1987), Discovery II (NASA/GRC, 2002), Human Outer Planet Exploration (NASA/MSFC, 2003), ICAN-II (The Pennsylvania State University)

The main advantage of fusion would be the very high specific impulse. Starships, powered by nuclear fusion reactions, should conceivably be able to reach speeds of the order of 10 % of the speed of light, based on energy considerations alone. In theory, a large number of stages could push a vehicle close to 30/40 % of the speed of light. In addition, a fusion rocket may produce less radiation than a fission rocket, reducing the mass needed for shielding. The main disadvantage would be the probable large mass of the reactor and the large mass of propellants (hydrogen and deuterium/tritium) that they should carry.

One theoretical spaceship designed to eliminate the need to carry large amount of fuel onboard is the ramjet spaceship. The interstellar ramjet collects interstellar matter to fuel its fusion rockets. As the ramjet moves, ions are caught by its magnetic field scoop which funnels them into a fusion reactor. Within the rocket, energy is released and fed back in some manner into the reaction products. These particles are then used to provide thrust for the vehicle. Most critics of this proposal correctly point out that the ramjet spaceship is not feasible for interstellar travel, since the amount of deuterium present in outer space is minimal.

Another possible way to overcome the problem of carrying large amount of fuel onboard the spaceship is to follow the suggestion of von Braun in his visionary book *The Mars Project*. Using the words of von Braun:

> In 1492, Columbus knew less about the far Atlantic than we do about the heavens, yet he chose not to sail with a flotilla of less than three ships, and history tends to prove that he might never had returned to Spanish shores with his report of discoveries had he trusted his fate to a single bottom. So it is for interplanetary exploration. […] The whole expeditionary personnel, together with inanimate objects for the fulfilment of their purpose, must be distributed throughout a flotilla of space vessels.

In the case of interstellar travel, spaceships carrying the necessary fuel can be sent in advance to specific points in space along the route to the target solar system for refuelling of the main spaceship. Although this would imply a slowdown of the expedition, since the main spaceship should decelerate and stop at the refuelling points, it has the advantage of requiring a lower mass ratio of the main spaceship and higher efficiency.

4.2.2 A Laser Driven Sail Starships

A time tested method of getting from place to place on the Earth is by sailing ship, with no needs for an engine because it gets its power directly from the wind and water. Similarly, a light sail starship carries no engine. Earlier designs for a light sail would use the light of and the high speed ejected gasses ejected from the sun. Some experiments have already been carried out. In 2004, a Japanese rocket has successfully deployed two small prototype solar sail into space. In February 2006, a 15 m solar sail was successfully put into orbit by a Japanese M-V rocket, although the sail opened incompletely.

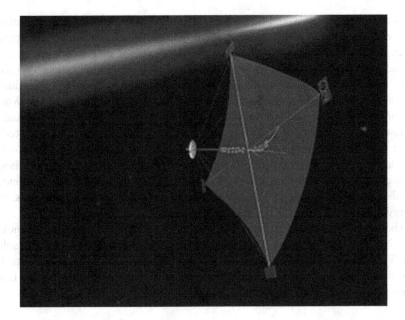

Fig. 4.5 Illustration of the unlit side of a half-kilometre solar sail (http://solarsails.jpl.nasa.gov.html)

Shortly after the invention of the laser in 1960, the concept of using beamed energy from lasers for interstellar propulsion appeared in a popular article by Forward.[8] At the time, Forward's ideas were considered impossible by physicists and engineers. However, further studies and advancement in laser technology have made this proposal acceptable and worthwhile of further consideration. The idea is to use the photon pressure of the laser light pushing against a giant sail to accelerate the interstellar probe up to relativistic velocities, which can be used for interstellar percussion missions and for manned missions later on (Fig. 4.5).

There is a light pushed sail concept that can. If high-powered lasers are used, in conjunction with a Fresnel zone lens, an energy beam can be focused at distances more than 40 light years away. In case of interstellar precursor missions, the light sail would have two sections. During the launch phase, the two sections would act as a single light sail. When the starship got near its destination the outer-most ring would disconnect, and act as a focusing mirror to focus the laser light onto the remaining section, thus decelerating the spaceship.

In case of the manned mission, the light sail would have 3 sections. During the launch phase, all three sections would act as a single light sail. Both the initial light sail and the Fresnel lens would be 1,000 km in diameter. The total weight of the Starship would be 80,000 t, including a 3,000 t payload. The power required to

---

[8] See Forward 1962.

accelerate such a craft would be 43,000 TW. This would allow 1/3 g acceleration. If accelerated for 1.6 years, the craft could reach a velocity of 50 % light speed. When the Starship got near its destination the outer-most ring would disconnect, and act as a focusing mirror to focus the laser light onto the remaining two sections, thus decelerating the spaceship. When it was time to go home, the middle section would disconnect and also perform the function of focusing mirror. When the ship got home, its small, remaining sail would be used to stop the craft at Earth.

To carry out a rendezvous mission with the direct laser photon propulsion technique would require very large (250 km) laser arrays, possibly in close solar orbit and drawing their energy requirements from the high solar flux available there. Although the laser arrays must be large to achieve the desired beaming distances and the energy must be high enough to push the probe to relativistic velocities, the energy flux from the laser is not high.

The advantage of such propulsion system is that it could use existing technologies. No new laws of physic would have to be discovered to create such a solar sail. The main disadvantage is the big engineering and economic problems to be solved in creating a sail hundreds of kilometres across energized by powerful laser beans placed around the sun or on the moon.

4.2.3 Antimatter Engines

The idea of using antimatter as the energy source in an interstellar vehicle has been in the technical literature for a long time. The annihilation of subatomic particles with their antimatter counterparts has the highest ratio of energy per unit of mass of any reaction known in physics. In the theoretical literature[9], two general classes of matter-antimatter propulsion systems have been explored: propulsion systems driven solely by the annihilation energy; and systems that use antiprotons to initiate a fission/fusion process in a compressed plasma or condensed material target.

The first class of system propulsion differentiate into: (1) those that directly use the products of antimatter annihilation for propulsion; (2) those that heat a working fluid/gas, which is then used for propulsion; and (3) those that heat a solid, high atomic number, refractory metal core. A propellant is then injected in the hot core and expanded through a nozzle to produce thrust.

The second class of matter-antimatter propulsion includes two concepts. In the first one, Antimatter-Catalyzed Micro-Fission/Fusion[10] (ACMF), a pellet of deuterium–tritium and uranium 238 is compressed with particle beans and irradiated with low intensity bean of antiprotons. The antiprotons are absorbed by the uranium 238 and initiate a fission process that heats and ignites the deuterium/tritium core, which expands and produces the pulsed thrust. In the second one, Antimatter-Initiated Microfusion (AIM), antiproton plasma is compressed by

---

[9] See Frisbee 2003.

[10] See Lewis et al. 1997.

electric and magnetic fields, while droplets of deuterium/tritium are injected into the plasma. The antiprotons annihilate in contact with the deuterium/tritium droplets, which heat the plasma to ignition conditions. The products of this ignition are directed to a magnetic nozzle to produce thrust (Fig. 4.6).

All of the concepts assume that the propulsion system would use half regular matter and half antimatter as fuel. When the regular matter and the antimatter are mixed together, complete conversion of the two into energy is attained in the form of neutrinos and gamma rays. The later easily materialize in high energy electrons and positrons. Approximately half of the energy is in the neutrinos which immediately escape from the system. Thus more than 50 % of the energy generated by annihilation can be converted either directly to thrust or to heat a working fluid that is then used to generate thrust.

The main differences between these different concepts are in the performance of each concept in terms of engine exhaust velocity, energy utilization efficiency and mass ratios. In the case of ACMF propulsion, the exhaust velocity can reach 100 km/s, while in the case of AIM propulsion; the exhaust velocity can reach 1,000 km/s. The amount of antimatter needed by a spaceship powered by an ACMF propulsion system with a payload up to 100 t for a one year round trip to Jupiter is about 10 micrograms (μg), while the amount of antimatter needed by a spaceship powered by an AIM propulsion system for a 50 years round trip to the Oort cloud is in the range of 100 μg.

The only antimatter concept that can reach 50/60 % of the speed of light is the one that uses the products of matter-antimatter annihilation directly to generate

Fig. 4.6 Conceptual image of a spaceship powered by a positron reactor (http://www.nasa.gov/images/content/145958main1_NTR_borowskii_smweb.jpg)

Fig. 4.7 Design of a high
capacity antimatter trap
(courtesy of NASA Marshal
Space Flight Center)

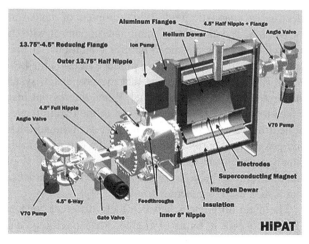

thrust, which would allow interstellar missions in a reasonable amount of time. One example of matter-antimatter annihilation propulsion is called the pion engine.[11] Pions are indirectly produced by proton-antiproton annihilation. The resulting charged pions will have a velocity of 0.94 of the speed of light, and a Lorentz factor of 2.93, which extends their lifespan enough to travel 2.6 m through the nozzle before decaying into muons.

60 % of the pions will have either a negative, or a positive electric charge. 40 % of the pions will be neutral. The neutral pions will decay immediately into gamma rays, which are useless for generating thrust. In realistic matter/antimatter reactions, collimation may not perfect, and some pions are reflected backwards by the nozzle. Thus, the effective exhaust velocity for the entire reaction drops to just in between 50 and 60 % of the speed of light. However, the antimatter mass needed to power these spaceships for interstellar travel is high. Depending on the payload of the spaceship it can range from few kilograms to few tonnes of antimatter.

Storage of antimatter is an important component of these types of spaceships.[12] Antimatter cannot be stored in a container made of ordinary matter because antimatter reacts with any matter it touches, annihilating itself and an equal amount of the container. Antimatter in the form of charged particles can be contained by a combination of electric and magnetic fields, in a device called a Penning trap.[13] This device cannot contain antimatter that consists of uncharged particles, for which atomic traps are used. In particular, such a trap may use the dipole moment (electric or magnetic) of the trapped particles (Fig. 4.7).

---

[11] See Schmidt et al. 1998.

[12] See Holzscheiter et al. 2000.

[13] Penning traps are devices for the storage of charged particles using a homogeneous static magnetic field and a spatially inhomogeneous static electric field. Currently Penning traps are used in many laboratories worldwide. For example, they are used at CERN to store antiprotons.

At high vacuum, the matter or antimatter particles can be trapped and cooled with slightly off-resonant laser radiation using a magneto-optical trap or magnetic trap. Small particles can also be suspended with optical tweezers, using a highly focused laser beam. Another, more hypothetical method is the storage of antiprotons inside fullerenes. In theory, the negatively charged antiprotons would repeal the electron cloud around the sphere of carbon, so they could not get near enough to the protons of the nuclei to annihilate with them.

4.3 Conclusions

Our analysis of propulsion systems indicates that nuclear fusion propulsion is the one in the most advanced stage of design development. Although this propulsion system can only reach about 10 % of the speed of light on energy consideration alone, it is possible that a multistage spaceship could reach higher fractions of the speed of light, making it capable of interstellar missions. The main problems with this propulsion system rest with the high mass ratio for these missions. It is our opinion that, on the basis of the existing literature, this propulsion system will be optimal for automated probe missions to the targeted planets for information gathering. The lower payload of the probes would allow less fuel, hence lowering the mass ratios of the probes.

The laser bean sail spaceships are the ones that present fewer technical problems, since they use existing technologies. This type of propulsion is favoured by many physicists, notably Freeman Dyson. Two main problems remain for the development of this propulsion system. The first one rests on the engineering and economics of building kilometres wide sails. Development of new materials and the possibility of assembling this spaceship in orbit may attenuate these problems. The second one rests on the technical and economic areas of building powerful arrays orbiting the Sun or on the Moon.

The matter-antimatter propulsion system remains the most attractive for interstellar mission, both manned and unmanned. It has the highest energy per unit of mass of any reaction, and it is capable to reach significant fractions of the speed of light. However, this propulsion system is in the least advanced stage of design development. Two main problems remain to be solved for this propulsion system to become feasible. One is technical and it is related to the storage of antimatter. Although Penning traps have been used successfully in laboratories, many problems have to be solved before an effective use on a spaceship. The second is related to the availability and costs of producing antimatter. This second problem will be extensively discussed in Chap. 5.

Box 4.1
Near Relativistic Velocities Equations

The specific impulse of relativistic rockets is the same as the effective exhaust velocity, despite the fact that the nonlinear relationship of velocity and momentum as well as the conversion of matter to energy has to be taken into account; the two effects cancel each other:

$$I_{sp} = v_e.$$

This is only valid if the engine does not have an external energy source (e.g. a laser beam from a space station; in this case the momentum carried by the laser beam also has to be taken into account). If all the energy to accelerate the fuel comes from an external source (and there is no additional momentum transfer), then the relationship between effective exhaust velocity and specific impulse is as follows:

$$I_{sp} = v_e / \left(\mathrm{sqr}\left(1 - \left(v^2/c^2\right)\right) \right) = \gamma_e v_e$$

where γ is the Lorentz factor.

In the case of no external energy source, the relationship between I_{sp} and the fraction of the fuel mass η which is converted into energy might also be of interest; assuming no losses,

$$\eta = 1 - \mathrm{sqr}\left(1 - \left(I_{sp}^2/c^2\right)\right) = 1 - \left(1/\gamma_{sp}\right).$$

The inverse relation is

$$I_{sp} = c \, \mathrm{sqr}\left(2\eta - \eta^2\right).$$

Here are some examples of fuels, the energy conversion fractions and the corresponding specific impulses (assuming no losses):

Fuel	H	I_{sp}/c
Matter-antimatter annihilation	1	1
Nuclear fusion	0.00712	0.119
Nuclear fission	0.001	0.04

In actual spaceship engines, there will be losses, lowering the specific impulse. In electron-positron annihilation, the gamma rays are emitted in a spherically symmetric fashion, and they almost cannot be reflected with current technology. Therefore, they cannot be directed towards the rear.

In order to make the calculations simpler, we assume that the acceleration is constant (in the spaceship's reference frame) during the acceleration phase; however, the result is nonetheless valid if the acceleration varies, as long as I_{sp} is constant.

In the no relativistic case, one knows from the (classical) Tsiolkovsky rocket equation that

$$\Delta v = I_{sp} \ln \left(m_0/m_1\right). \tag{4.1}$$

Assuming constant acceleration a, the time span t during which the acceleration takes place is

$$t = (I_{sp}/a) \ln (m_0/m_1).$$ (4.2)

In the relativistic case, the equation still valid if a is the acceleration in the spaceship's reference frame and t is the spaceship's proper time because at velocity 0 the relationship between force and acceleration is the same as in the classical case. Solving this equation for the ratio of initial mass to final mass gives

$$(m_0/m_1) = \exp (at/I_{sp})$$ (4.3)

where "exp" denotes the exponential function. Another related equation gives the mass ratio in terms of the end velocity Δv relative to the rest frame (i.e. the frame of the rocket before the acceleration phase):

$$(m_0/m_1) = [(1 + (\Delta v/c))/(1 - (\Delta v/c))]^{(c/2\,Isp)}.$$ (4.4)

For constant acceleration, $\Delta v/c = \tanh [at/c]$ (with a and t again measured on board the rocket), so substituting this equation into the previous one and using the identity $\tanh x = (e^{2x} - 1)/(e^{2x} + 1)$ returns the earlier Eq. (4.3).

By applying the Lorentz transformation on the acceleration, one can calculate the end velocity Δv as a function of the spaceship frame acceleration and the rest frame time t'; the result is

$$\Delta v = at' / \left(sqr \left(1 + [(at')^2/c^2] \right) \right).$$ (4.5)

The time in the rest frame relates to the proper time by the following equation:

$$t' = (c/a) \sinh [(at/c)].$$ (4.6)

Substituting the proper time from the Tsiolkovsky equation and substituting the resulting rest frame time in the expression for Δv, one gets the desired formula:

$$\Delta v = c \tanh [(I_{sp}/c) \ln (m_0/m_1)].$$ (4.7)

References

Alcubierre, A.: The warp drive: Hyper-fast travel within general relativity. Class. Quantum Gravity 11(5), L73–L77 (1994)

Bignami, G.F., et al.: Project 242: Fission fragments direct heating for space propulsion—program, synthesis and applications to space exploration. Astro. Acta (2011)

Bussard, R.W.: Galactic matter and interstellar flight. Astro. Acta 6, 179–194 (1960)

Bussard, R.W., DeLauer, R.D.: Nuclear Rocket Propulsion. McGraw-Hill, New York (1958)

Clark, R.A., Sheldon R.B.: Dusty Plasma Based Fission Fragment Nuclear Reactor. Joint Propulsion Conference and Exhibit, July 10–13 (2005)

Forward, R.L.: Pluto: The gateway to the Stars. Missiles and Rockets **10**, 26–28 (1962)

Freeman, D.: Interstellar transport. Physics Today 41–45 (1968)

Frisbee, R.: How to build an antimatter rocket for interstellar missions. In: Huntsville, A.L. (ed.) AIAA Paper 4676, 39th Conference (2003)

Holzscheiter, M.H., et al.: Production and Trapping of Antimatter for Space Propulsion Applications. Laboratory for Elementary Particle Science, Pennsylvania State University (2000)

Lewis, R.A., et al.: Antiproton-Catalyzed Microfission/Fusion Space Propulsion Systems for Exploration of the Outer Solar System and Beyond. In: El-Genk, M. (ed.) Space Technology and Applications International Forum, 26–30 Jan 1997, Albuquerque NM, CONF-970115, American Institute of Physics, New York (1997)

Matloff, G.L., Chiu, H.H.: Some aspects of thermonuclear propulsion. J. Astron. Sci. **18**, 57–62 (1970)

Norem, P.C.: Interstellar Travel: A Round Trip Propulsion System with Relativistic Velocity Capabilities. Am. Astron. Soc. **69**, 388 (1969)

Oberth, H.: Die Raketen zu den Planetenraumen [The Rocket into Planetary Space]. Nürnberg Uni-Verlag (1964)

Schmidt, G.R., et al.: Antimatter Production for Near-term Propulsion Applications. NASA Marshall Space Flight Center Paper (1998)

Thorne, S.K.: Black Holes and Time Warps: Einstein's Outrageous Legacy. Norton New York (1994)

Chapter 5
Financing of Space Exploration and Colonization

There is nothing neither so big nor so crazy that one out of a million technological societies may not feel itself driven to do, provided it is physically possible.

Freeman Dyson

Our scenario foresees an early phase of development of propulsion technologies with a 10 year period of manned mission definition studies, automated probe payload definition studies and development efforts on critical propulsion technology areas. Once the automated probe design is finalized, robotic space exploration of targeted planets for the collection of the necessary information for future colonization will be launched. Given the distance to nearby stars, this phase should be completed in 30/40 years. Development of man-rated propulsion systems would continue while awaiting the return of the automated probe data. Assuming positive data returns from the probes, a manned exploration starship would be launched.

This chapter will explore the costs of a program of interstellar exploration and colonization, the ways in which it could be financed, and the correspondingly roles of the public and private sectors. Financing of such an enterprise would not be easy without the mobilization of both public and private resources. We envisage that the public sector will be heavily involved in the construction and management of a space-port, where spaceships will be assembled and launched, and in the robotic space exploration of targeted planets. We envisage that the private sector will participate significantly in the exploration and colonization of targeted planets.

These distinctions follow a well established principle of economics. Purchases of goods and services intended to create future benefits, such as infrastructure investment or research spending, are carried out by public entities once the demand of the general public for such goods and services arises. In our scenario, this demand will arise at the time of the discovery of a habitable planet in the neighbourhood of the solar system, such as the ones recently discovered around nearby stars Tau Ceti, Epsilon Eridani and Alfa Centaury. Private markets instead will allocate goods and services more efficiently in the later phase of interstellar exploration and colonization, once the goals of the project are better defined.

Given the actual and medium term budgetary situations of major countries around the world, finding public resources for the creation of the space

G. F. Bignami and A. Sommariva, *A Scenario for Interstellar Exploration and Its Financing*, SpringerBriefs in Space Development, DOI: 10.1007/978-88-470-5337-3_5,

Table 5.1 Military expenditures and non military space programs expenditures in 2012 (in billion US$)

	2011	30 years extrapolation	%
World military expenditures	*1,617.7*	*48,531*	*100*
USA military expenditures	739.3	22,179	45.7
Rest of the world military expenditures	878.4	26,352	54.3
Total space expenditures	*24.9*	*748.0*	*100*
NASA[1]	16.0	480.0	64.2
ESA	5.6	168.0	22.5
International space station	3.3	100.0	13.4
Space expenditures as a % of military expenditures			
NASA[1]	1.0	1.3	
ESA	0.3	0.3	
International space station	0.2	0.2	

[1]Excluding expenditures for the ISS
Source Military expenditures: IISS; space budgets: NASA and ESA

infrastructures and the automated interstellar probe exploration will not be easy. A realistic possibility of finding these resources, without disrupting major civilian government programs, rests with a gradual reduction in military expenditures through disarmament agreements concerning both nuclear and conventional weapons.

At present, nation states are spending a large amount of resources for defence purposes (see Table 5.1). In 2011, the United States alone spent about 740 billion US$, while the combined military expenditures of the rest of the world were estimated at about 878 billion US$ per year. Total world military expenditures amount at about 1.6 trillion US$ per year. Extrapolating over 30 years, this is about 48 trillion US$.

Comparing actual expenditures for peaceful space exploration with military expenditures, the first ones appear minimal. It is estimated that the total development and operational costs of the International Space Station have been about 200 billion US$ over a 30 years period or less than one half of a percent of the projected military expenditures over the same period. NASA budget for 2011, excluding the contributions to the International Space Station, is about 16 billion US$, while ESA budget for 2011 is estimated at 5.6 billion US$. Extrapolating NASA and ESA budgets for 30 years, it gives a total of 748[1] billion US$, which represent about 1.6 % of total projected military expenditures. If one includes the estimated cost of the International Space Station, this would represent 1.8 % of total projected military expenditures. We have no reliable data from other countries, but it is estimated that these would be minor with respect to the budgets of NASA and ESA, except maybe for the unknown space budget of China.

---

[1] The above comparison is to be taken with caution because of the exclusion of space related expenditures of other countries as Japan and China.

Fig. 5.1 NASA-ESA 2011
budget (NASA and ESA)

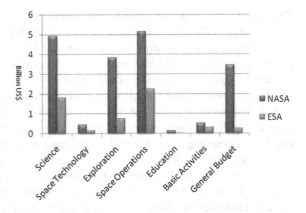

A closer look at the details of the Budgets of NASA and ESA indicates also
that the development of new propulsion technologies, whether for manned solar
system or interstellar exploration, is a minor objective for these organizations. This
is apparent when considering the details of the expenditures of NASA and ESA
(see Fig. 5.1). The resources allocated to the development of space technologies
are about 1 % of total expenditures of NASA and ESA.

It is evident that a future endeavour of interstellar exploration and colonization
requires a civilization in which human beings see themselves as inhabitants of a
single planet and in which global governance is conducted on a cooperative basis.
The key question is whether our present civilization is ready to embrace a culture
of cooperation and non violence, gradually reducing military expenditures and
devoting resources to space exploration. This question will be taken up in Chap. 6.

Our goal in this chapter is to develop a sustainable model for persistent, long-
term, private/public sectors investment into the myriad of disciplines needed to
make long-distance space travel viable. In other words, we shall try to develop an
economic-financial vehicle providing sustained investment over a long time horizon, together with the agility to respond to the accelerating pace of technological
change. We will first try to estimate the costs involved in a program of interstellar exploration and colonization, and then we will put forward some suggestions
of the role of governments and the private sector in the financing of deep space
exploration and colonization.

5.1 Estimated Costs of a Program for Interstellar
Exploration and Colonization

A long term program for interstellar exploration and colonization would require
three stages. The first stage would involve the design of a propulsion technology
adequate for manned solar system and automated probe interstellar missions.

The second stage would involve the construction of a space transportation infrastructure, allowing initially solar system exploration and automated probe interstellar missions. Waiting for the return of information from the automated probes, this phase would also include the development of a starship capable of reaching a significant fraction of the speed of light. A third stage would involve manned exploration and colonization of nearby solar systems.

It is important to point out that an interstellar exploration program does not need a plan. It will come without planning. Our developing capabilities in controlling fusion, beaming laser energy through space, generating and storing antimatter, tailoring new structural materials, and reducing the size and cost of computers take place as a result of general research progress, and will make interstellar exploration technologically feasible. Resources are already dedicated to the developments of these technologies independently from an interstellar exploration program, because of the growing demand for clean energy, the awareness by centres generating antimatter of the potential demand for it, the need for new materials for energy conservation, and the demand for better and cheaper communication, and storage and faster elaboration of data in a modern society. As W. von Braun once said:

> Only a miraculous insight could have enabled the scientists of the 18th century to foresee the birth of electrical engineering in the 19th century. It would have required a revelation of equal inspiration for a scientist of the 19th century to foresee the nuclear power plants of the 20th century. No doubt, the 21st century will hold equal surprises. But not everything will be a surprise. It seems certain that the 21st century will be the century of scientific and commercial activities in outer space, of manned interplanetary flight, and the establishment of permanent human footholds outside the planet Earth.

All that will be required to focus these technologies on interstellar exploration will be the political will. If you ask the men who built the moon ship what was the main factor behind the Apollo program, they might give you only one answer. They would say that the main ingredient behind their adventure was leadership, someone throwing down the gauntlet and committing the country to a single clearly defined goal: man, moon, and decade.

5.1.1 First Stage

The main goal of the first stage is to develop propulsion technologies adequate for high speed manned solar system and automated probe interstellar missions, and to solve various physics and engineering problems related to these missions within 10 years. The costs of this development program per se should not exceed few billion US$ over a period of 10 years.

Propulsion technologies. In Chap. 4, we indicated that three propulsion technologies are theoretically capable of reaching velocities within a significant fraction of the speed of light: nuclear fusion, laser bean, and matter-antimatter annihilation based technologies. Research should be done on all these technologies in order to select an optimal one. Optimality should be defined in terms of costs and technological feasibility.

At the moment, we do not know which of the techniques proposed for controlled nuclear fusion will prove feasible for the propulsion portion of the interstellar exploration mission. Advancement in the next years of nuclear fusion for power generation here on Earth should give valuable insights on which direction we should go. Moreover, the problem of the efficient utilization of the different forms of energy generated during nuclear fusion has yet to be solved, since only 20 % of the released energy will be contained in the kinetic energy of the fusion particles. This should also be the subject of analysis and proposals.

As far as the matter-antimatter annihilation based propulsion, different antimatter propulsion concepts have been analyzed in the theoretical literature, including systems driven solely by annihilation energy and systems driven by antimatter initiated fusion. Only systems driven solely by annihilation energy can reach speed close to 40/50 % of the speed of light allowing a human visit to a nearby star in a reasonable amount of time. Systems driven by antimatter initiated fusion can reach a maximum velocity of only 10^3 km/s, allowing manned solar system exploration and automated probe interstellar missions.

In Chap. 4, we pointed out that the amount of antimatter required to power propulsion systems driven by annihilation energy is very high, in the range of kg and tonnes, while the amount of antimatter required by system driven by antimatter initiated fusion for solar exploration and interstellar precursor missions is much lower, in the rage of 1–100 µg. The main problems rest with the production capacity of antimatter and its costs. At present, the only facilities producing antimatter are CERN in Geneva and the Fermi Lab in Chicago. The production of antimatter was added at these facilities during the late '70s and '90s to increase the energies and yield of particle collision experiments. The annual yields of antimatter is equivalent to few nano grams of antimatter, which is 3–4 orders of magnitude less than the ones required for mission using propulsion based on antimatter initiated fusion.

The actual costs of producing antimatter on a gram scale or above are staggering high. A recent study indicates that the costs for producing a gram of antiproton are in the range of 63 trillion US$. However, historical precedents indicate that the costs of new materials, such as liquid hydrogen and Uranium 235, were extremely expensive at first, but dropped substantially once the production infrastructures were in place and improvements were implemented. Recent studies have also indicated that the efficiency of antimatter production can be improved substantially by optimizing proton acceleration energy and incorporating improved collection methods. It is estimated that, with such upgrades, the costs for 1–100 µg of antimatter—quantities required for solar system and automated probe interstellar missions—could drop to 60 million US$. If these costs are attained, they would make these missions economically feasible.

To reach this objective, the amount of money to be invested for the upgrading of present facilities producing antimatter is in the range of 20–30 billion US$. It is important to stress that this money would probably not come out of the budget for the development of solar system and automated probe interstellar propulsion technologies. Rather, it could well come out of the budgets of the centres producing antimatter once they realize the growing demand for it, as demand for antimatter may arise not only by the prospective demand for interstellar exploration, but also for medical and other related uses.

Engineering. As far as the engineering problems related to automated probes interstellar mission, they are several. Although the designs of automated probes like the Voyagers, are well understood, initial missions envisioned for extra-solar planetary probes present new aspects that have yet to be worked out. The new probes must be fully autonomous and resilient due to the exceptionally long mission lifetimes and distances from Earth, and the completely mysterious environment they will find at their destination. The rigors and length of a journey involving high gravity accelerations with high energy density engines, the years of bombardment against interstellar matter, and the decades of operation means that a new level of self-diagnostic, self-repairing, multiply-redundant probe design must be developed. The autonomous features of the computational circuitry of the probe will have to be developed to the extent that the probe will exhibit semi-intelligent behaviour when presented with new and unforeseen circumstances. We note with optimism that already four space probes, launched 40 years ago, are still happily travelling in interstellar space.

5.1.2 Second Stage

The second phase will be the most expensive part of the interstellar exploration program. It will involve the construction of a space-port for the assembling and launching of automated probe interstellar and manned solar system missions, and the development of propulsion systems for fast interstellar travel. How much will it cost to complete a space transportation infrastructure and the construction and launch of space probes?

It is difficult, at this time, to give an estimate of these costs. The only available estimate is the cost of construction of a new space-port station where spaceships could be assembled and lunched. Based on data from the International Space Station, it is likely that the cost of building and operating the new space station in low earth orbit would amount to about 200/300 billion US$ over a period of 30 years.

The key virtue of orbital assembly is that it eliminates the tight connection between the size of the expedition and the size of the rockets used to launch it. That has three big advantages, one obvious and the others more subtle. The obvious advantage is that the rockets do not need to be big enough to launch the entire expedition all at once. A more subtle advantage is that there is no need to fix the exact size of the expedition when one chooses the size of the rocket. It is hard to estimate the mass of a big complex system correctly at the very start. In particular, if the expedition ends up being a little heavier than expected, one can just launch a few more rockets to assemble it. Whereas if one plans on launching everything on one big rocket, and it turns out to be too small, this would be a real problem.

Fig. 5.2   Chesley Bonestell painting (Chesley Bonestell Foundation)

Back in the 1950s, it was taken for granted that an expedition to the Moon would be assembled in Earth orbit. In 1946, Wernher von Braun[2] prophesied the construction of space stations in orbit. The design was made very familiar to the American public at the First Symposium on Space Flight on 12 October 1951 in New York City. The design was popularized in the series in Colliers magazine, illustrated with Chesley Bonestell painting (Fig. 5.2).

In 1969, Von Braun proposed the construction of a space station for the assembly and launch of spaceships to a Mars mission, which was not approved by the US Senate because of the lack of government funds due to the heavy involvement of the United States in the Vietnam War.

Later on, the idea of a space station was re-proposed and realized in cooperation with five participating space agencies: NASA, the Russian Federal Space Agency, JAXA, ESA, and CSA. The ownership and use of the space station is established by intergovernmental treaties and agreements. According to the original Memorandum of Understanding between NASA and RSA, the International Space Station was intended to be a laboratory, observatory and factory in space. It was also planned to provide transportation, maintenance, and act as a staging base for possible future missions to the Moon, Mars and asteroids. Unfortunately, the International Space Station is on an orbit which is too inclined on the ecliptic to be used as a spaceport and a staging base for future interplanetary and interstellar missions. Hence, a new one has to be built for the purpose of deep space exploration.

It is important to decide whether is will be convenient to build the space port in Earth orbit or in Lagrangian point between the Earth and the Moon. The main disadvantage of a space port in Earth orbit is that the effects of the non-Keplerian forces, i.e. the deviations of the gravitational force of the Earth from that of a homogeneous sphere, gravitational forces from Sun/Moon, solar radiation pressure and air-drag must be counteracted with a rather large expense of fuel.

The main advantage of a space port located at a Lagrangian point is the avoidance of counteracting action to keep the stability of the orbit of the space station.

---

[2] See von Braun 1952 and 1991.

Lagrangian points are locations in space where gravitational forces and the orbital motion of a body balance each other. They were discovered by French mathematician Louis Lagrange in 1772 in his gravitational studies of the "Three body problem". An object placed at a Lagrange point will be in a state of gravitational equilibrium, and will orbit with the same period as the bodies in the system. In other words, in the Earth−Moon system, an object in a Lagrange point will keep pace with the Moon in its orbit about Earth.

There are five Lagrangian points in the Earth−Moon system. Among these five Lagrangian points, two are stable points (L4 and L5), while L1, L2 and L3 are unstable. The forces of gravity and orbital motion are precisely balanced at these points, but even a slight nudge will send any object at them drifting off. Because of this instability, most attention has been given to the two stable points L4 and L5, located in the Moon's orbit but off the position of the Moon (see Fig. 5.3). These positions have been studied as possible sites for a space port for the assembling, launch and repair centre for starships moving throughout the Solar System and beyond. It is difficult to estimate how much would it costs to build a space station in one of the two Lagrangian points, since there is no precedent in building such a station. One can only say that it would most likely costs much more than building a port station in low earth orbit.

Services to and from this space port station could be carried out by the new emerging private companies, such as SpaceX and Virginia-based Orbital Sciences Corporation. Space Exploration Technologies Corporation, or SpaceX, is a space transport company headquartered in Hawthorne, California. It has developed the Falcon 1 and Falcon 9 launch vehicles, both of which were designed from conception to eventually become reusable. SpaceX also developed the Dragon spacecraft to be flown into orbit by the Falcon 9 launch vehicle, initially transporting cargo and later planned to carry humans. On 25 May 2012, SpaceX made history as the world's first privately held company to send a cargo payload, carried on the Dragon spacecraft, to the International Space Station. Other private companies may be formed in the coming years, which would reduce costs of low orbit transportation, ensure the sustainability of the new space port and enable new exploration (Fig. 5.4).

Fig. 5.3 Earth−Moon
Lagrangian points (http://
www.ottisoft.com)

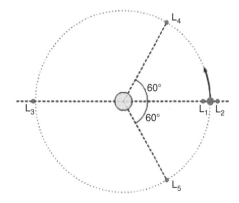

Fig. 5.4 View from the international space station of the SpaceX Dragon spaceship as the station's robotic arm moves Dragon into place for attachment to the station (http://www.citizensinspace.org)

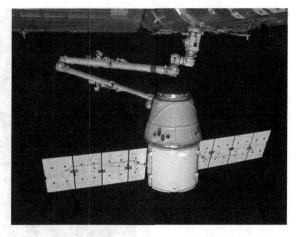

The costs of automated probes building, assembling, and launching is more uncertain. We have the historical data on the total costs of the American outer planets probe, Voyager 1 and Voyager 2. The cost of the Voyager 1 and 2 missions, including the spaceship development, launch, and mission operations through the Neptune encounter, was 865 million US$. However, the interstellar probes will be significantly different from the Voyagers, particularly in terms of mass, payload, and propulsion system. The mass of the Voyagers was 800 kg with a span of 13 m. Power was provided to the spaceship systems and instruments through the use of three radioisotope thermoelectric generators.

The mass of the interstellar probes will be substantially bigger because of the different payload and propulsion system. The payload should include long-range sensor capabilities to gather the data from orbit. These, in turn, require very high resolution capability from orbital altitudes, which drive up the size of the transmitting and collecting apertures desired. Development efforts on multiple function emitters, antennae, sensors and data processors are needed to minimize the payload (Fig. 5.5).

The propulsion system based either on nuclear fusion or on antimatter initiated fusion will also add to the mass of the probe. Lastly, sophisticated computer systems, should be installed on the spaceship capable of reasoning about its own state and the state of the environment in order to predict and avoid hazardous conditions, recover from internal failures, and ultimately meet the science objectives despite the uncertainties. Taking into consideration the higher costs of constructing such a spaceship and the estimated cost of the antimatter need for this long journey, it is safe to say that the costs will exceed 10 billion US$ for each probe sent in its exploration mission.

Costs of manned solar system exploration can be estimated with greater accuracy. According to R. Zubrin, the costs of a very conservative Mars exploration program should be of the order of 25–50 billion US$.[3] Costs of an exploration program

---

[3] See Zubrin 1996.

Fig. 5.5 Voyager 2 (NASA)

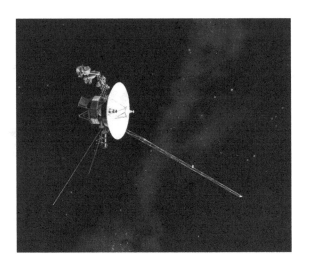

to the asteroid belt have not been estimated yet. However, we can safely assume that it will be in the same order of magnitude as the costs of the Mars exploration program. Assuming the launch of several probes and manned exploration missions to Mars and the asteroids, and the costs of constructing the space-port, it is not unreasonable to estimate that the total costs of these missions will be in the range of 400/500 billion US$ over a period of 30 years. This is not an unreasonable figure. It is estimated that a reduction of 1.5 % of global defence spending over 30 years would allow government funds for the exploration of the solar system and interstellar precursor missions, including the building of a space port in a low Earth orbit.

5.1.3 Third Stage

The third stage would involve the construction and launch of a manned spaceship based on matter-antimatter annihilation energy or nuclear fusion. At present, it is impossible to advance an estimate of the costs of such a mission. We can safely say that the costs of constructing a spaceship based on matter–antimatter annihilation energy would be in the range of 30–40 billion US$.

The major cost would be the cost of antimatter, which will determine the feasibility of such a mission. The costs of producing larger quantities of antimatter, required for interstellar missions, would remain very high regardless of the extent to which efficiency of producing it is improved. In order to reduce the costs for large quantities of antimatter production, the power utility costs should drop substantially, which would occur only if energy from nuclear fusion would become available.

It is expected that nuclear fusion will be developed first for energy production on Earth. The present and future energy needs required by worldwide economic development and environmental concerns about the use of fossil fuels and nuclear fission are powerful motivations for the development of clean energy sources. Global energy demand is expected to double by 2050, as the world's population soars and more people gain access to energy. Today, the supply of energy is mainly based on fossil fuels and nuclear fission. Presently there are no shortages of fossil fuels. Whether this will remain so in the future depends on the potential peaking of oil production. What is certain is that a more intensive use of fossil fuels can put further pressures on the ecosystem, and more intense use of nuclear fission is problematic given the present political climate and the real problem of disposing of the radioactive waste.

Clean energy technology could come to the rescue of the increasing hunger for energy due to worldwide economic development. Among possible best candidates for clean energy generation, nuclear fusion is a key technology, since nuclear fusion could be the biggest carbon-free utility-scale energy source, with much reduced problems of radioactive waste disposal.

Numerous advanced experiments of nuclear fusion are presently under development. There are three promising experiments[4] in nuclear fusion: (1) a stellarator is a device used to confine hot plasma with magnetic fields in order to sustain a controlled nuclear fusion reaction; (2) a tokomak is a device using a magnetic field to confine plasma in the shape of a torus; (3) laser-based inertial confinement fusion is a device using powerful lasers to heat and compress a small amount of hydrogen fuel to the point where nuclear fusion reactions take place.

According to the latest report of the International Atomic Energy Agency (IAEA), magnetic fusion machines are not expected to produce net energy until 2019, and electricity generation until 2027. The situation for inertial confinement nuclear fusion machines is similar, but with longer development times. Although it is difficult to define the development costs of nuclear fusion technologies, current analysis indicates that the estimated costs for the development of nuclear fusion are in the range of several billions US\$. If these earlier experiments are successful, the costs of nuclear fusion reactors will go down because of economies of scale.

The deuterium and lithium used to fuel our fusion power plants are inexpensive and plentiful. The combined cost of these fuels is less than \$0.0001/kWh, more than 100 times lower than coal, natural gas, or uranium. Lithium is abundant and widely available. The current annual lithium production is 16,000 Mt with 28.5 Mt of known land reserves and 250 Gt of seawater reserves. If fusion power plants were used to generate all of today's electricity, land and sea reserves of lithium would be sufficient for 207 million years of production. Deuterium is easily extracted from water and the world's seawater constitutes a reserve of

---

[4] Some important modern stellarator experiments are Wendelstein 7-X, in Germany, and the Large Helical Device, in Japan; (1) a tokamak is a device using a magnetic field to confine plasma in the shape of a torus. Important experiments are: Tore Supra, at the CEA, Cadarache, France in operation since 1988; FTU, in Frascati, Italy in operation since 1990; and Alcator C-Mod, MIT, Cambridge, USA in operation since 1992; (2) The National Ignition Facility, or NIF, is a large, laser-based inertial confinement fusion research device located at the Lawrence Livermore National Laboratory in Livermore, California.

23.3×10^{12} Mt of deuterium. If fusion power plants were used to generate all of today's electricity, this seawater reserve would last for 66.6 billion years.

Moreover, fusion technology does not incur any costs associated with carbon dioxide sequestration or disposal, which will likely be factored into future electricity costs derived from coal and other hydro-carbon-based power plants. Fusion technology also does not incur significant costs of disposing highly active long-lived radioactive wastes associated with fission plants. It is thus reasonable to expect that within the next 20–30 years, the utility costs of electricity could be lowered to such an extent that would allow the production of antimatter in quantities necessary for manned interstellar travels.

5.2 Role of Governments and the Private Sector in the Development of Interstellar Propulsion Technologies

The first stage starts with a 10 year period of mission definition studies, automated probe payload definition studies and development efforts on critical technology areas. We believe that this early phase should be guided by two basic principles: (1) the organization in charge of this research program should be free from external influences, depending entirely on the skills, initiative and devotion of its members; and (2) the goal of the research program should be to design a spaceship with the objective of minimizing the costs of interstellar exploration. To this end, the various propulsion systems indicated in Chap. 3 suitable for interstellar exploration should be analyzed competitively in order to reach the costs objective.

The organization in charge of the first phase should have the task of promoting industry/academic/government coordination in the development of propulsion technologies. We propose that a non profit organization is set up initially under the supervision of the United Nations. Alternatively, a United Nations Agency, like the Committee on Space Research (COSPAR) founded in 1958, the same year as the establishment of NASA, could be considered as a candidate organization. Membership to this organization should be enlarged to space agencies, academics, and enlightened wealthy individuals and entrepreneurs. Funding for this early phase should be based on large scale donors and endowment contributors and crowd funding. Large scale donors should include wealthy private individuals, private foundations and space agencies, as these last ones are the depository of the knowledge on space technology.

The discovery of a habitable planet would rekindle the interest of individuals capable of donating significant sums of money, which are considered a primary source of seed capital for building an endowment. Mobile and information technologies have enabled fundraising via small contributions from many individuals. Crowd funding (or crowd sourced capital) has been demonstrated already for political campaigns and disaster relief efforts. A compelling crowd funding program offers two opportunities: a source of seed money and a sense of ownership in the contributing public's mind. The sense of ownership could provide support for future efforts.

Grants, scholarships, prizes, and contracts can be awarded to worthwhile undertakings, in science, engineering, humanities, and the arts, in pursuit of attaining interstellar flight. The endowment should be an incubator for innovative ideas regarding the conception, development and operation of a class of spaceship that has unprecedented resilience, i.e., the ability to achieve envisioned objectives even if the spaceship performance and/or the environment are not as expected.

5.3 Role of Governments and the Private Sector in Automated Probe Interstellar and Manned Solar System Missions

The goal of this phase of interstellar space exploration is to build, assemble, and launch: (1) probes to the targeted planet for the collection of the necessary information for manned interstellar exploration and colonization; and (2) manned solar planetary exploration missions.

5.3.1 Interstellar Probes

Construction of the space-port is the first step for the automated probe interstellar missions. How long will it take to build this space-port? The only evidence that we have is the timing of the construction of the International Space Station. It took 12 years to complete this station. The station has been under construction since November of 1998. In that year the first piece of its structure, the Zarya Control Module, was launched into orbit with a Russian Proton rocket. In 2008, the two-billion-dollar science lab Columbus was added to the station, increasing the structure to eight rooms. The International Space Station was completed by 2010, allowing six crew members to live and work in a space larger than a typical five-bedroom house.

In our scenario, the construction of the interstellar probes will take place during the building period of the space-port. We then assume that the phase of interstellar probe exploration to nearby planets within 5–15 light years from our solar system would take from 15 to 30 years to be completed. Even if we had a propulsion system that could get our probe up to the speed of light instantaneously, it would still take 4.3 years for the probe to travel to the nearest star, and 4.3 years for the messages from the probe to travel back. If a probe accelerates at one earth gravity (1 g) for 1 year, it will have reached greater than 90 % of the speed of light in a distance of 1/2 light years. It could now coast for 3 years, and then decelerate for 1 year to arrive at Alpha Centauri in about 5 years. This time is not much longer than the minimum travel time of 4.3 years. It will take then another 4.3 years for the information to reach Earth. For other interesting stars, like Tau Ceti at 11.8 light years, the minimum mission time is 24 years.

We propose that the construction and management the new space port, and the assembling and launch of interstellar probe/s should be carried out by an international organization, like COSPAR, under the supervisory authority of the United Nations. Funding of this venture should come from the reduction of military expenditures as described in Sect. 4.1.2. This international organization should be autonomous in setting up a program for interstellar probe exploration and in designing the appropriate procurement strategies, while being accountable to the United Nations. As von Braun said in his visionary study on the Mars Project:

> Since the actual development of the long range liquid rocket, it has been apparent that true space travel […] can only be attained by the coordinated might of scientists, technicians and organizers belonging to very nearly every branch of modern science and industry. Astronomers, physicians, mathematicians, engineers, physicists, and test pilots are essentials; but no less so are economists, businessmen, diplomats and a host of others. We space racketeers of all nations have made it our business to rally these kind of talents, which, in the nature of things, is synonymous with the future of rocketry.

5.3.2 Solar System Exploration

Waiting for the information return from the interstellar probes, exploration of the solar system is warranted. We propose that two missions should be carried out: one to Mars and the other to the asteroid belt. The mission to Mars is important because: (1) it is a place where humans might discover some form of primitive life, increasing our understanding of how life originates; and (2) geological surveys of Mars can identify deposits of raw materials necessary here on Earth. Manned missions to the asteroid belt are important because analysis of the chemical and mineral composition of the asteroids could lead to a better understanding of the origin of the solar system and, possibly, to the extraction of these minerals for uses here on Earth.

Future exploration missions into the solar system are also critical for testing the new propulsion systems and humans capabilities that are unique and critical. We propose that the propulsion system for these missions is based on antimatter initiated fusion. One of the advantages of using antimatter fusion propulsion systems is to reach speed up to 1,000 km/s, shortening the time travel of the manned mission. As an example, a human round-trip mission to Jupiter travelling at 1,000 km/s with a 10–100 metric ton payload can be completed in 1 year, requiring an antimatter quantity of 1 to 10 μg.

These missions will test human capabilities such as their physical adaptability in long term missions in space, intellectual flexibility, and capacity for innovation in operations. In particular, we can understand how human will respond more flexibly and intelligently to new or unexpected features of the region they are exploring They will have also the ability to inspire Earth's population, as these exploratory voyages have the potential—within our lifetimes—to answer age-old questions about how life begins, whether life exists elsewhere, and how humans will exist in the future.

Funding of these explorations should be done by a mix of private and public money, as those who pay for such missions are more likely to be interested in benefits for themselves (search for raw materials and minerals) or for the human race as a whole (scientific research, and understanding the origin of life).

5.4 Role of the Private Sector in Interstellar Manned Exploration and Colonization

Once all relevant information on the targeted planet are received back to Earth from the space probes, the launch of manned interstellar space exploration and colonization could start. We maintain that manned interstellar exploration and colonization should be funded by the private sector. Historical comparison comes here to our rescue: the early colonization of North America. The original impetus of North America colonization was economic and strategic. In the 16th century, trades with the riches of the Americas were monopolized by Spain and Portugal. The British government and the merchants in London were looking for settlements in North America as a base for breaking the monopoly of Spain and Portugal and for attacks on the Spanish treasure fleets. Later, a variety of motives would inspire English colonies in North America: trade, fishing, escaping religious persecutions, and the desire to start life over as part of a new society.

Since the passage to North America was costly and most of the colonists could not afford to sustain these costs, their travel was funded by a group of merchants in England in exchange for a share in the profits of the new colonies. The establishment of the colonies was carried out in different ways. A number of English colonies were established under a system of independent Proprietary Governors, who were appointed under mercantile charters to English joint stock companies to fund and run settlements. Most notable was the Virginia Company, which created the first successful English settlement at Jamestown and a second settlement at St. George's, Bermuda.

Noticeable was the case of the colony of Massachusetts. The Pilgrim Fathers were a small group of English Puritans who had separated from the Anglican Church. Most of these Separatists were farmers, poorly educated and without social or political standing. In 1617, discouraged by economic difficulties and their inability to secure civil autonomy, the congregation voted to emigrate to America. Unable to finance the costs of the emigration with their own meagre resources, they negotiated a financial agreement with Thomas Weston, a prominent London iron merchant. To finance their journey and settlement the Pilgrims had organized a joint-stock venture. Capital was provided by a group of London businessmen who expected to profit from the colony.

After some delays and disputes, the voyagers regrouped at Plymouth aboard the 180-t Mayflower. It began its historic voyage on Sept. 16 1620, with about 102 passengers in the cramped spaces of the ship. The ship came in sight of Cape Cod on November 19. The colonists had been granted territory in Virginia but the Mayflower turned back and dropped anchor at Provincetown on November 21.

That day 41 men signed the so-called Mayflower Compact, a "plantation covenant" modelled after a Separatist church covenant, by which they agreed to establish a "Civil Body Politic" and to be bound by its laws. The compact became the basis of government in the Plymouth Colony.

During the first winter, more than half of the settlers died as a result of poor nutrition and inadequate housing. Without good harbours or extensive tracts of fertile land, it became a colony of subsistence farming on small private holdings once the original communal labour system was ended in 1623. In 1627 eight Pilgrim leaders assumed the settlement's obligations to the investors in exchange for a 6-year monopoly of the fur trade and offshore fishing (Fig. 5.6).

Private funding of the early North America colonization allowed for the mobilization of the necessary resources for sustaining such ventures, particularly in view of the fact that the finances of the British government were stretched to the limit in building a fighting ocean fleet and in the wars on the continent. Given the present and foreseeable budgetary constraints of national governments, it is not far from the truth to say that private funding of interstellar exploration and colonization will be an important element in the implementation of interstellar exploration and colonization.

In the proposed a scenario, it is assumed that the spaceship accelerates at one earth gravity (1 g) for 1 year, which will allow the spaceship to reach about 90 % of the speed of light in a distance of ½ light years. It would then coast for 10 years and decelerates for 1 year to reach destination. The round trip would then be 24 years, including the time allowing for acceleration and deceleration of the spaceship, since the spaceship could not accelerate/decelerate to/from its cruising speed instantaneously.

Time dilation, the slower passage of time for the adventures on board of the spaceship with respect to Earth inhabitants would facilitate the financing of space

Fig. 5.6 Model of the ship that carried the Pilgrims to America (http://www.whereto gotravelusa.com)

exploration. According to Einstein's time stretching factor[5], the time for covering the round trip will be about 24 years for the perspective of people on board of the spaceship, while from the perspective of the inhabitants of Earth it will be 71 years.[6]

We assume that a group of wealthy people (the merchants) decide to fund the voyage to an identified inhabitable planet by forming a company, Deep Space Company (henceforth called DSC). We further assume that DSC is willing to cover 25 % of the cost of building the spaceship and the fuel needed in exchange from profits from the new colony, while the colonists will supply all the materials necessary for the colonization. We assume that the colonists will sign an agreement with the merchants granting monopoly of trade on natural resources for a limited period of time in exchange for the passage to the new planet. We also assume the colonists will sign a "Covenant" establishing the rules on how to run the new colony, which will form the basis for the government of the colony.

Assuming that the costs of the spaceship and the fuel is 4 US$ (the number of zeros after 4 is yet unknown), the DSC will enter into an agreement with a bank to deposit 1 US$ (the number of zeros after 1 is yet unknown) at the bank for the duration of the round trip of the spaceship, earning a 2 % yearly interest rate in real term. Straightforward calculation of the accumulated compounded interest rate will give a value of the deposit at the end of the 71st year of 74.78 US$. DSC then decide to issue a zero coupon bond with a duration of 71 years.

A zero-coupon bond (also called a discount bond or deep discount bond) is a bond bought at a price lower than its face value, with the face value repaid at the time of maturity. It does not make periodic interest payments, or have so-called "coupons," hence the term zero-coupon bond. DSC decides that the price of the zero coupon bond is 4 US$ (the number of zeros after 4 is yet unknown) and its face value at maturity is 64.8 US$, implying a yearly discount factor of 4 %. It has to be noted that, in this arrangement, the zero coupon bond is default risk free, since it is guaranteed by the deposit of DSC at the bank.

There is a large market for these types of bonds. Pension funds and insurance companies like to own long maturity zero-coupon bonds because of the bonds' high duration. This high duration means that these bonds' prices are particularly sensitive to changes in the interest rate, and therefore offset, or immunize the interest rate risk of these firms' long-term liabilities. The use of these instruments is aided also by an anomaly in the tax system of some countries, which allows for deduction of the discount on bonds relative to their par value. This rule ignores the compounding of interest, and leads to significant tax-savings when the interest is high or the security has long maturity.

At the return of the spaceship to Earth, the zero coupon bond will be paid out and DSC will own the spaceship, which in time will constitute part of the fleet of spaceships servicing the commerce and transportation between the new colony and Earth. It is thus conceivable that this financial structure or similar ones will

---

[5] Einstein 1905.

[6] The γ time stretching factor is equal to 3.1622.

succeed in funding the space exploration and colonization by wealthy private individuals and groups of colonists. It is also realistic to assume that, as population expand in the new colony; the colonists will establish new colonies in other close by star systems. Like the ancient Polynesians populated the Pacific Ocean leapfrogging island by island, it is thus possible to imagine that, in the centuries to come, humanity will "green" the Milky Way galaxy, increasing the chances of survival of the human race.

5.5 Conclusions

In this chapter we have analyzed a plausible model for private/public investments in space exploration and colonization. The positive element behind this scenario is the large economic benefits that will occur initially here on Earth. As demonstrated by the Apollo program and the International Space Station, the economic benefits of these programs are big. The economic impact of the five-year gear-up of the Apollo program is well documented. More than 400,000 highly skilled jobs were created in industry. The technologies transferred from space to the national economy were largely responsible for whatever increases in productivity there were in industry, commerce, and the home, for over 20 years.

Perhaps chief among them was the integrated circuit. Its development for the space program led indirectly to the proliferation of the home computer 20 years later and to a technical revolution, which scientists call "Moore's Law", holding that the number of transistors on any one integrated circuit is being halved in size every 2 years. In other words, cell phones, computers, and all kinds of communications devices are getting progressively smaller, yet more powerful, adding to the increases in productivity of our economies. Developing nations saw that their lives could be dramatically improved through the use of satellites for communications, Earth remote sensing, health, and education.

The reduction of military expenditures necessary for the development of the space programs outlined in this chapter will most likely have a zero effect on economic activities, as a large part of the industries involved in military production, such as the aero-space industries, will certainly participate in the new space programs with minimal effects on employment and growth. The only uncertainty about this scenario is our present level of civilization, as our scenario implies a level of civilization based on cooperation and non violence. This argument will be the subject of the last chapter of this book.

Box 5.1
Costs of Producing Antimatter

An article by G. R. Schmidt et al. (1998) has proposed a way to calculate the energy costs of producing antimatter. The creation of antimatter is an energy intensive process. The conversion of input energy E_{in} into rest mass energy of the collected antiprotons E_{out} can be expressed in terms of efficiency:

$$\eta = E_{out}/E_{in}. \tag{5.1}$$

Because of the equivalence between E_{out} and antiproton mass M_a ($E_{out} = M_a c^2$), the energy required to create a unit mass of antiproton is:

$$E_{in}/M_a = c^2/\eta. \tag{5.2}$$

This energy requirement is inversely proportional to the conversion efficiency since the speed of light is constant. Thus the antiproton can be at most one half of the total rest mass produced in a perfectly efficient conversion process. The total energy cost C of producing antiproton is obtained by multiplying Eq. (5.2) by the power utility cost C_{grid} and M_a:

$$C = C_{grid}M_a c^2/\eta. \tag{5.3}$$

Equation (5.3) indicates that the conversion efficiency factor η is a major factor in determining the energy costs of producing antimatter. In present day facilities, this factor is around 4×10^{-8}. Assuming a power utility cost of 0.1 US\$ per kilowatt hour, this results in an energy costs of about 63 trillion US\$ per gram of antiproton. However, studies have shown that the efficiency of production of antiprotons can be improved substantially by optimizing proton acceleration energy and by incorporating improved collection methods. This could result in a 3 to 4 order of magnitude improvement over current capabilities.

The above results indicate that, even by improving the efficiency of production, the costs of production of 1 gram or above of antimatter are still very high. However, the prospects of applications involving small amount of antimatter are more promising. With upgrades in the efficiency of production of antimatter, the energy costs of producing 1–100 μg of antimatter required for solar system and interstellar precursor missions would be in the range of 60 million US\$, which would make such missions economically feasible. Interstellar exploration missions, requiring kg or ton of antimatter, would become economically feasible only in the case of substantial decline of the power utility costs, which may happen only with the generation of electricity by nuclear fusion.

References

Einstein, A.: On the electrodynamics of moving bodies. Annalen der Physik **322**(6), 132–148 (1905)

Schmidt, G.R., et al.: Antimatter production for near-term propulsion applications. NASA marshall space flight center paper, 1998

Von Braun, W.: Das Marsprojekt, Studie einer Interplanitariscen Expedition. Bechtle Verlag, Esslingen (1952)

Von Braun, W.: The Mars project. University of Illinois Press, Urbana (1991)

Zubrin, R.: The case for Mars: the plan to settle the Red Planet and why we must. Touchstone, New York (1996)

Chapter 6
The Uncertainties

> *To what purpose should I trouble myself in searching out the secrets of the stars, having death and slavery continuously before my eyes?*
>
> Attributed to Pythagoras

This book has developed a scenario for interstellar exploration and colonization, which is hypothetical but not unrealistic. It has identified the forces that may shape the future in this direction. This chapter will analyze the uncertainties surrounding this scenario that may push the future in different directions. The development of the technologies for interstellar exploration requires money and time, with money initially coming from the public purse.

As pointed out in Chap. 5, budgetary constraints of major countries around the world make it difficult in the short-medium term to find public money necessary for the investments in a program for interstellar exploration, unless global military expenditures are reduced. It is estimated that a reduction between 1 % of global defence spending over 30 years would allow not only the breakthrough in interstellar propulsion technologies, but also the exploration and colonization of the planets of the solar system, leading the way to interstellar exploration and colonization within the next 100 years.

Future interstellar exploration and colonization thus requires a civilization in which human beings see themselves as inhabitants of a single planet and in which international relationships are based on a cooperation rather than conflict. The key question is thus whether our present civilization is ready to embrace a culture of non violence in the management of the international relationships, gradually reducing military expenditures and devoting resources to space exploration without disrupting other major government programs. Hence critical uncertainties for the present scenario are political and cultural.

The political literature has analyzed two ways in which an international system can organize itself on the basis of cooperation, mutual respect and non violence: (1) the establishment of some kind of world government with the monopoly of military power; or (2) the achievement of a stable division of the world into independent states or group of states with armed forces strictly confined to the mission of defending their own territory. Although the distinction between offensive and defensive

G. F. Bignami and A. Sommariva, *A Scenario for Interstellar Exploration and Its Financing*, SpringerBriefs in Space Development, DOI: 10.1007/978-88-470-5337-3_6, © The Author(s) 2013

uses of weapons is difficult to make and will be subject to many controversies, the second alternative seems more realistic, at least in the short/medium term.

What will take to achieve a stable world order among independent nation states? Following H. Bull[1], at least four conditions have to be considered to achieve such stable world order. The first is a consensus on common interests and values. The consensus on common interests and values would provide the foundation for the second condition, a set of rules accepted by all nations for peaceful coexistence. The third is justice. The fourth condition is a cosmopolitan culture. It is important that this cosmopolitan culture is shared not only among the elites but also among larger strata of the population.

6.1 Historical Lessons

This section will analyze past mistakes and successes as humanity struggled to achieve fairer and more tolerant societies. Three historical periods will be discussed: the crisis of the classical world, the development of the global economy in the later part of the 18th century and the 19th century, and the events in the 20th century.

6.1.1 The Classical World Crisis

The Classical world crisis—the fall of the brilliant scientific civilization centered in the Library of Alexandria—provides a good example of how a society failed to address the question of justice, particularly slavery, leading to the demise of a civilization that had accomplished spectacular successes in the areas of science and of the understanding of the natural world.

In 332 BC, Alexander the Great founded the city of Alexandria. Ptolemy, the successor of Alexander, transformed the city into one of the greatest centres of learning in the Greek world. According to the earliest source of information, the library was initially organized by Demetrius of Phaleron, a student of Aristotle, under the reign of Ptolemy I. The Library at Alexandria was charged with collecting the entire world's knowledge, and most of the staff was occupied with the task of translating works onto papyrus paper. Immigrants from elsewhere in the eastern Mediterranean turned this relatively new place into a great cosmopolitan centre. It is probably here that the word cosmopolitan achieved its true meaning.

At the Library, the best minds of the ancient world established the foundation for a systematic study of mathematics, physics, astronomy, biology, geography, medicine and literature. Euclid produced a textbook on geometry on which we still

---

[1] See Bull 1977.

learn today. Eratosthenes calculated with great accuracy the size of the Earth, the distance between the Earth and the Moon, and the distance between Earth and the Sun. Hipparchus anticipated that the stars are born and eventually perish. He went on to catalogue the positions and magnitudes of the stars to detect such changes.

Aristarchus introduced a theory in which the Earth rotates daily and revolves annually around the Sun. Heron published a well recognized description of a steam-powered device (Fig. 6.1).

Among his most famous inventions was a wind wheel, constituting the earliest instance of wind harnessing on land. In optics, Hero also formulated the Principle of the Shortest Path of Light: if a ray of light propagates from point A to point B within the same medium, the path-length followed is the shortest possible.

Although we have mentioned only a few of the brilliant minds of the Library of Alexandria, it is clear that there were the seeds of the modern world. However, there was a notable missing element in the works of the Library. As far as we know, there is no record that these brilliant minds ever challenged the political, economic and religious foundation of their society. They questioned everything in the fields of science, but never questioned the justice of slavery. The abundance of cheap slave labour was an important factor in the inability of the classical world in recognizing the potential of technology to free people. As a consequence science and learning remained the preserve of privileged few, and most of their discoveries in the field of mechanics and steam technology were used to advance weaponry, to encourage superstition, and to the amusement of kings.

It is not controversial that slavery weakened the vitality of the Classical world. Although there were ways to escape slavery under the patronage of influential roman families, the majority of slaves had no hope to change their status in society. The advent of Christianity changed all that. Under the rules of the new religion, a convert became a member of the community through baptism without any perceived taint from his previous life and social status. In the early Christian communities, persons of low social status were allowed to increase their stand in the

Fig. 6.1 Illustration of
Hero steam power device
(http://explorable.com/
images/aeolipile-illustration.gif)

community through acts of devotion and martyrdom. These communities also took care of the widows, orphans and the destitute. Encouraging solidarity provided also a good reason for a convert to remain loyal to the community.

In the early years, Christian communities became a magnet for slaves and persons of low social status. Later on, when Constantine declared his allegiance to the new faith, persons of higher social status, particularly in the Eastern part of the Empire, started to convert to Christianity, which slowly became the official religion of the Empire. As the Christian Church was consolidating his power and attempting to eradicate the pagan influence, many Christian leaders, particularly Cyril, the Archbishop of Alexandria, largely identified science and learning with paganism. This brought the final act of the tragedy when the Christian mob came to burn the Library and there was nobody to stop them.

This is magnificently described in Amenabar's movie *Agora*, centring on the life and death of Hypatia. She was a mathematician, astronomer, physicists and the head of the Neoplatonic School of philosophy, an extraordinary range of accomplishments for a single individual. The movie describes the increasing isolation from the imperial institutions of this island of learning, which was surrounded by an ocean of ignorance and hate.

The burning of the Library and the subsequent killing of Hypatia are portrayed as in the classical Greek drama. It was as an entire civilization was erased and most of its ideas, discoveries, and passions were lost forever. The western world went through 1,000 years of darkness before the ideas of the Library were discovered again during the Renaissance, powering the advent of the modern world (Fig. 6.2).

Fig. 6.2 Library of Alexandria (http://www.deskarati.com)

6.1.2 The Mid-18th and 19th Centuries

The Enlightenment provides a good example of how a set of ideas developed in England, particularly in Scotland[2], provided the foundation for a consensus around values and interests among European nations in the late 18th and 19th centuries, which led to the integration of the world economy. The moral philosophers of the 18th century sought to mobilize the power of reason in order to reform society and advance knowledge. Their works were the source of critical ideas, such as the centrality of freedom, democracy, the rule of law and reason as primary values of society as opposed to the divine right of kings or traditions, which explains their opposition to hierarchy, inequality, including racial inequality, slavery, and empire. They stressed the need of a private realm of society independent from the state, leading to increased society's autonomy and self-awareness, along with creating an increasing need for the exchange of information.

The value system advanced by the moral philosophers provided the lens to see a world shaped by technology, capitalism and democracy, and the individuals' struggle to find a place in it. They motivated a number of people in Scotland and in England to advance science and reform their society. James Watt's invention created the work engine for the Industrial Revolution. Charles Darwin developed his theory of biological evolution, strongly influenced by the works of Charles Leyell and James Hutton on geology, which augmented the age of the Earth to millions of years from the commonly accepted Biblical age of 6,000 years. Darwin work was the foundation of the subsequent developments of biology. James Maxwell scientific breakthrough on electromagnetism provided the conditions for the Second Industrial Revolution, driven initially by electricity.

James Watt's invention did more than providing the work engine for the Industrial Revolution It also created the basis for the industrialization of transportation. Breakthrough for steamship allowed the opening of new and fast routes for international trade, while railways provided fast and cheap ways for the transportation of goods and raw materials within Europe. The steam engine also freed entrepreneurs in their choice of the location of factories. Previously, entrepreneurs built their factories by a river or in the proximity of raw materials sources, such as coal mines. The steam engine allowed them to choose the location of their business where it suited them. This generally meant a place close to the route for the supply of the raw materials and the transport of their products, and where they could find cheap and ready supply of labour, which in turn usually meant a city. In other words, the steam engine made industrial production essentially an urban activity.

The industrial cities of the 19th century were born: Manchester, Liverpool, Birmingham and Glasgow. Urban population expanded very rapidly, attracting rural immigrants in search for work. The growing population exceeded the capacity of the cities to offer affordable and safe housing, and adequate sewerage and sanitation. The horrors of these early industrial cities are well documented in the

---

[2] See Chitnis 1976 and Stewart 1991.

work of Dickens. Squeezed between squalid living conditions and low wages, workers fought back, as manifested in the general strike in Glasgow in 1820. These battles foreshadowed the future of the relation between labour and capital in the following 100 years. There was however, a particular distinction of the labour battles in England. Workers were demanding not only decent living conditions and higher wages, but also a sense of individual dignity and independence. They did not want a revolution, but to be part of progress.

During the 19th century, middle class liberalism in England learned to extend the benefits of civilizations to those left behind. Politically, middle class liberalism set in motion events and trends that would ultimately bring the vote to the working class and heal old wounds. It pioneered a schooling system extended to all segment of the population, which in time became the basis for a meritocratic society. Due to advances in modern medicine, particularly in the field of public health to halt dangerous epidemic diseases, numerous reforms were introduced to increase sanitary conditions in workplaces (Fig. 6.3).

The economic, scientific, political and social successes of the Industrial Revolution in England gradually influenced other European countries west of Russia. During the 19th century, countries like Germany, the Austrian Empire, Belgium, and Italy gradually adopted the British system. The process of adaptation was frequently incomplete, as some of the basic conditions present in England were missing in those countries. These countries were still rural societies and the political system was mostly influenced by aristocratic, agrarian landlords. The middle class was still relatively small and unable to push substantial reforms in

Fig. 6.3 Charles Dickens' Characters and Glasgow Slum (http://www.arnelbanawa.com and http://en.wikipedia.org/wiki/File:Slum_in_Glasgow,_1871.jpg)

their societies. As a consequence, many of the continental European countries did not develop into genuine democracies.

As these countries lacked the extensive network of economic relationships that England enjoyed due to the build up of a world empire in the late 18th century and early 19th century, the lead centre in shaping international relationships remained in London. This facilitated the creation of a free trade area in Western Europe. When England repealed most of its duties on trade in 1846, all Western European countries followed and by 1870 Western Europe had become a free trade zone.

Outside Europe, the establishment of global free trade required political and military interventions. England, by far the largest colonial empire, established free trade in its colonies. However, other non European empires and states were not ready to abandon their traditional system as in the case of the Ottoman Empire, China and other Asian countries. Political and military pressures were exercised on them to force them to adopt free trade, as in the case of the Opium War with China. Other countries, particularly in Latin America, were weak and frequently succumbed to political pressures.

The only notable exception was Japan, which adopted self reforms following the Meiji Restoration in 1868. Japan adopted institutions such as the military, police, government administration, and universities according to western models, although retaining its core institutions of social life such as the family. By the turn of the century, Japan was becoming the first modern state in Asia. However, Japan did not become a genuine democracy, but was the first constitutional nation in Asia.

The economic, political and military pre-eminence of England and the adaptation of rules of international relationships by European countries, particularly free trade, allowed a century of relatively peaceful relationships in Europe west of Russia. This was achieved because of a consensus of interests of newly industrialized nations on the rules of international relationships. Industrialization came gradually in these nations, so it was in their interest during this period to abide to London, which was the main centre of finance and the leading industrialized nation. However, there was no consensus on values, as many of these nations were unable to adopt the values of a liberal society.

6.1.3 The 20th Century

The 20th century was characterized by the disintegration of the world economy, the end of Europe's domination of the world, the emergence of the United States as the major world power, and, in the latter part of the century, to the reintegration of the world economy. Moreover, the decolonization after the 1950s marked the appearance on the world stage of new countries with claims for a redistribution of wealth and resources, and the demand for justice.

The first 50 years. The end of the 19th century saw new developments in international relationships in Europe and in the United Sates. Two factors were at the origin of these changes. The *first* one was the reaction of the losers of the globalization process. German farmers threatened by cheap imports of grains and meat, and American farmers threatened by the depress price cause by the gold standard in the mid 1870s pushed their governments to raise tariffs. With the exception of England, most countries reverted to protectionism in 1878.

The *second* one was the changed relationship between industries and governments in most western continental countries, most notably Germany. Industries strongly pushed governments to protect their domestic markets (infant industry theory) and to sustain their efforts in gaining market shares worldwide. Political elites saw an opportunity to enhance the legitimacy of the state by protecting domestic interests. All the above was in marked contrast with the notion of the role of government, which had been developed in the previous 100 years.

The politicization of trade led initially to a race to divide up the remaining parts of the world that had not being colonized, generating tensions among rival European nations and the Unites States. It was followed by the aggressiveness of German export industries in gaining export market shares. From the mid-1890s on, the German government devoted significant economic resources to building up a navy competing with the British Royal Navy for world naval supremacy. The growing aggressiveness of the German government led to a new system of alliance between England, France, Russia and the United States. According to some authors, the aggressiveness of German industries for export markets, coupled with that of the German government, were the main causes of the First World War, which was the beginning of the end of Europe domination of the world and shook profoundly the international economy.

It is well known that the economic and financial consequences of the First World War were at the origin of the financial crisis of the end of the 1920s, the disintegration of the world trade system, and the great depression of the 1930s. Politically, World War I radically altered the political map, with the defeat of the Central Powers and the 1917 Bolshevik seizure of power in Russia. Meanwhile, existing victorious Allies such as France, Belgium, Italy, Greece and Romania gained territories, while new states were created out of the collapse of Austria-Hungary, and the Russian and Ottoman Empires.

Despite the pacifist movements in the aftermath of the war, the territorial losses caused irredentist and revanchist nationalism to become important in a number of European states. Irredentism and revanchism were particularly strong in Germany because of the significant territorial, colonial, and financial losses incurred by the Treaty of Versailles. However, the rise of revanchist nationalism in Germany did not affect the elections, as democratic governments were elected throughout the 1920s.

The Weimar Republic was parliamentary democracy established in 1919 in Germany to replace the imperial form of government. The ensuing period of liberal democracy continued up to the early 1930s. In its 14 years, the Weimar Republic faced numerous problems, including hyperinflation, political extremisms, and hostility from the victors of the First World War. The democratic governments overcame many of the requirements of the Treaty of Versailles, reformed the

Fig. 6.4 *Cartoon* 1932—our
last hope—Hitler (http://www.
pages.uoregon.ed)

currency, and unified tax politics and the railway system, as well as having a
unique cultural impact on art, music and cinema.

What pushed the balance towards political extremists were the events between
1929 and 1932. In 1929, monetary policy developments in the United States triggered a virtual halt of the country capital outflows and an economic recession.[3] As
Germany was not able to finance the rising government expenditures and the repa-
ration repayments without inflow of capital from abroad, developments in the
United States triggered a recession in Germany.

Despite the downturn of the international business cycle in 1930 and the deteri-
oration of the balance of payment current account, the German government introduced deflationary policies, because of fear of inflation. These policies aggravated
the downturn of the German economy, causing a rapid increase in unemployment,
which reached the unprecedented level of 14.3 % at the end of 1931.

Deflationary policies were reversed in mid 1932 by the new government led by
Van Papen. Despite the fact that the last democratic government had adopted the
appropriate monetary and fiscal policies in 1932, the social cement of the country
had crumbled under the severe recession of 1929–1932 and the Nazi Party gained
power. The aggressiveness of the German foreign policies in the following years
culminated in the Second World War, which ended Europe's domination of the
World (Fig. 6.4).

---

[3] See Field 1984 and Meltzer 1976.

The second 50 years. The end of the Second World War saw Europe in ruin, the rise of the United Sates and the Soviet Union as the major international powers, and the division of the world in two blocks. An "Iron Curtain" divided Europe and Germany, while China became part of the communist bloc. This led the United States to mobilize its vast resources to revive the economies of Western Europe and Japan. It sustained political movements demanding democratic institutions in the former enemy countries. It favoured the formation of an economic community among Western European countries as an escape from the extreme forms of nationalism that had devastated the continent in the past.

Relationships among Western countries were characterized by an associative multilateralism, which gave considerable political leeway to the different countries in many areas, with the exception of defence. The legal and institutional framework for a free world economy was established at Bretton Woods in 1944. The basic principles were fixed exchange rates, free movement of goods and capital, and freedom of governments to pursue economic policies. The basic idea was to ensure that international cooperation remained compatible with full employment policy. To this end, several multilateral institutions were created: The World Bank, the International Monetary Fund and the General Agreement on Tariffs and Trade. The principles established at Bretton Woods were applied gradually and never entirely, as they were curtailed whenever a country felt that this was necessary for domestic stability and to protect their autonomy in internal policy making.

Defence was the only area organized under the hegemonic leadership of the United States. Common fear of communism created the consensus of Western Europe, with the exception of France, and of Japan to accept the hegemony of the United States in the defence area. The United States set up a series of regional defence alliances around the world directed to contain potential threats from the Soviet Union.

The Soviet bloc was primarily a military security zone in which the Stalinist system of society and economy were imposed by the Soviet Union. The relationships between the Soviet Union and its allies were thus hegemonic in all areas. The exceptions were Yugoslavia and China. Both countries, although remaining communist regimes, loosened the grip of the Soviet Union. Yugoslavia tried to establish relationships with the rest of the world, while China isolated itself completely from the rest of the world until the end of the 1970s.

Economic relationships between the two blocs were mostly in terms of barter, although in the 1970s exchange between the rubbles and the US$ was established. Politically, the relationships between the two superpowers were initially dominated by military matters as both possessed nuclear weapons and their deployment was a major component of their military strategies. After the crisis in Berlin (1958) and Cuba (1962) when the use of nuclear weapons were seriously considered, the two superpowers accepted the notion of coexistence on the basis of mutual deterrence. The competition between the two superpowers was then in the fields of economic growth, social development, technology and space competition. Military conflicts were relegated to peripheral areas (such as Vietnam and Afghanistan), but never involved a direct confrontation of the two superpowers.

In the aftermath of the Second World War, the ideas of liberty, justice and the rights of individuals for self determination were extended gradually to the rest of the world, through the decolonization process. The process of decolonization was supported by the two superpowers with economic and military aid to the new countries that choose the "right" side. As the new states shared a number of common interests, including the need to overcome underdevelopment and not to be horded by the two superpowers, they formed a movement of bloc free nations. This marked the rise of the demand from the ex-colonial countries for a redistribution of wealth and resources, and an ending of their dependence and subordination in which they stranded with respect to Western countries. The main avenue for expressing their interests was the United Nations where the new countries gradually achieved the majority in the General Assembly.

Three major events dominated the international relationships in the period from 1970 to 1992. The *first* one occurred in 1971 when the United States unilaterally pulled out of the Bretton Woods Accord taking the US off the Gold Exchange Standard and allowing the dollar to "float". The *second* one was the oil shock of 1973, when the members of Organization of Arab Petroleum Exporting Countries or the OAPEC, consisting of the Arab members of OPEC plus Egypt, Syria and Tunisia, proclaimed an oil embargo. The *third* was the opening up of China to the world.

The first event was the result of the contradictions of the Bretton Wood basic principles and the weakening of the budgetary position of the United States because of the Vietnam War. After 1945, American companies started to build plants in Europe. At first, this was due to the inability of European nations to pay imported goods in dollars. After the establishment of the Common Market, the tariff imposed by it encouraged American companies to continue with that trend. The foreign profits of the American multinational corporations were the primary source of the Eurodollar market in London.

The growth of this market allowed these capitals to evade all form of government control and weakened the ability of central banks to control interest rates, money supply and capital movements. The rapid increase of the United States debt at the time of the Vietnam War resulted in pressures on the dollar until the Federal Reserve gave up and pull out of the Gold Exchange Standard and floated the dollar. The creation of unregulated international capital markets had the longer term effect of weakening the capabilities of governments' policies of influencing economic developments in their countries.

The second event, the oil shock, allowed a group of developing countries to reacquire control of a vital commodity. The increases in the price of oil reversed the traditional flow of capital, as the oil exporting nation's accumulated vast wealth. Some of the income was dispensed in the form of aid to other underdeveloped nations whose economies had been caught between higher prices of oil and lower prices for their own export commodities and raw materials amid shrinking Western demand for their goods. Much was absorbed in massive arms purchases that exacerbated political tensions, particularly in the Middle East. On an international level, the price increases of petroleum disrupted market systems by changing competitive positions. At the macro level, economic problems consisted of both inflationary and deflationary impacts of domestic economies. Politically, high oil prices, disrupted supply, and recession, created a strong rift within NATO.

The third event was the opening of China to the world. In 1978, the Chinese government embarked on a policy of opening to the outside world in a planned way and step by step. China adopted policies oriented to developing the foreign-oriented economy, generating foreign exchanges through exporting products and importing advanced technologies. These developments attracted the strong interest of transnational corporations, from car industry to electronic industries, which were looking for new bases to diversify their global production network to areas where there was an abundance of skilled and cheap labour. Slowly, China became a major hub in their global production network and a major exporter of manufactured goods. Politically, China reasserted its independence from Russia, which had always been the traditional "enemy" together with Japan. These developments allowed the United States to operate a triangular policy, putting pressures on the Soviet Union in the attainment of its global objectives.

6.2 The Present State of International Relationships

The end of the Cold War marks the beginning of the most recent phase of globalization, characterized by an integration of the world economy that was experienced only at the end of the 19th century. The major differences between the two periods are: the appearance of new nations on the world scene characterized by different stages of their industrialization; the altered balance of power between nation-states and markets due to unregulated capital markets; the widening income gaps both within and among nation states; and the changed global power structure established by the Cold War.

6.2.1 Globalization

The rapid progress in technology of information and communication and the reintegration of the world economy have created the conditions for many human activities—such as arts, science, commerce and finance—to be global. People engaged in these activities are able to meet anywhere in the world and consider themselves as inhabitants of the planet. Political elites, both in the developing and the developed countries, have engaged in multilateral agreements for the establishment of free trade and in scientific cooperation for the development of new clean energy sources and space research. In some regions of the world, the cooperation process has gone deeper, involving the subordination of cherished national sovereignties to common goals. The French, the Germans and other European countries are mostly advanced in subordinating their sovereignties, for which they used to fight so dearly and deadly in the past, to common goals, institutionalized in the European Union.

However, there is a large part of the world population who has been excluded from the benefits of the globalization process. Economic globalization has contributed to lift many people out of poverty, but there have been many losers in this process, particularly in the developed world. The economic globalization has been accompanied by a widening of the gap between poor and rich in the developed world. A 2011 report by the Organization for Economic Co-operation and Development (OECD) concluded that income inequality in OECD countries is at its highest level for the past half century. The average income of the richest 10 % of the population is about nine times that of the poorest 10 % across the OECD, up from seven times 25 years ago. In both Israel and the United States inequality has increased further from already high levels. Other traditionally more egalitarian countries, such as Germany, Denmark and Sweden, have seen the gap between rich and poor expand from 5 to 1 in the 1980s to 6–1 today.

The report indicated that there are many reasons for economic inequality. Among the factors affecting economic inequality, the report indicated: (1) globalization, as multinational corporations have used sophisticated legal and financial means to circumvent the bounds of local laws and standards, in order to leverage the labour and services of unequally developed regions against each; (2) technological changes, (3) policy reforms, particularly taxes. In a progressive tax system, the level of the top tax rate will have a direct impact on the level of inequality within a society, either increasing it or decreasing it. The lowering of the level of the top tax rate in the US is seen by many economists as an important cause of the widening income gap in that country.

The widening income gap within our nations has been accompanied by a lesser ability of nation-states to influence economic development and to access their resources, particularly tax revenues, seems to have been impaired because of the growing importance of the unregulated, international capital markets. Regionally, particularly in the European Union, the provisions of the welfare state have also been threatened, thereby diminishing the legitimacy of nation-states in the eyes of their citizens.

The gap between poor and rich nations remains wide. According to estimates of the United Nations, 66 % of world GDP is concentrated in Europe and North America, with Asia scoring 24 %. Africa is scoring particularly badly with only 1 % of world GDP. Asian countries, at the exclusion of Japan, are in a more advanced state of industrialization and integration of their economies in the global network of transnational corporations. In the less developed countries, critical gaps are: (1) human resource development gap; (2) technology gap; (3) knowledge and information gaps; (4) production gap; (5) trade and terms of trade gaps.

The widening and persistent economic gap between the "haves" and the "haves not" and the perception of losing traditional identities have pushed many "haves not" in the arms of radical religious and/or ideological groups both in developed and developing countries. We are used to hear in the media about Muslim fundamentalism, but there is a surge in developing countries of an equally dangerous Christian fundamentalism (Fig. 6.5).

Fig. 6.5 Christian
fundamentalist comic (*Atheist
Cartoons Archive*) (http://www.
atheistcartoons.com)

These movements uphold a literal reading of the Bible and extremely conservative social values, particularly on the role of women and abortion. These movements are eradicated in the Anglo-Saxon world, particularly the United States, but are also spreading in some continental European countries. In the later 20th century, fundamentalists made use of television as a medium for evangelizing and became vocal in politics as the Christian Right.

In developed countries, the growing uncertainties about the economic future for young people, white and blue collars workers, and small entrepreneurs may well be one of the main causes of the emergence and of the growing importance of xenophobic political parties. Many of the "haves not" have embraced the radical ideology of the no global movements, which see in globalization the cause of every evils in our societies. The rise of fundamentalisms and radically different cultures has been accompanied by a surge in violence in the pursuit of their objectives and by the rejection of long cherished democratic values in some of the countries targeted by this violence.

6.2.2 Geopolitical Developments

Globalization has also made the world power structure more diffuse. Since the early 1990s, the world seemed to have become unipolar with the emergence of the United States as the dominant power. Because of the perceived threats from transnational terrorism, failed states, and international crime, the Unites States have been engaged in several regional conflicts. Kosovo, Afghanistan and Iraq conflicts and Latin America fight of the drug traffic are examples of the global military involvement of the Unites States. This has led to a strong increase in government expenditures, particularly military expenditures. At the same time, several US administrations have succumbed to the pressures of the economic and financial elites to reduce taxes. Increase in government expenditures and reduced revenues led to massive government deficits and increases in government debt in the US.

For a while, this unstable situation has been managed by the increased absorption of the US government debt by emerging countries, like China. By acquiring dollar denominated bonds, these countries maintained the exchange rate of their currencies with respect to the US$ and hence the competitiveness of their products in the eyes of American consumers. At the same time, loose US monetary policies sustained private consumption and economic growth. This has led to the appearance of the twin debt in the US, both internal and external, and to several bubbles in financial markets. The latest break up of the real estate bubble in the US led to the economic crisis in the US and the slowdown of economic growth worldwide. Politically, this has induced a shift in global balance of power in the direction of emerging nations, particularly China, which however continue to depend on the US imports for the their growth.

At present, these developments seem to lead to cooperation between the US and the emerging countries in the solution of the global economic problems. However, history tells us that in the decade prior to 1914 many observers thought that economic interdependence[4] between the UK and Germany were making war between these two nations unlikely, if not impossible. However, the war came. In the months following the stock market crash in October 1929, many observers thought that the US was facing a recession. However, the inability of the UK and the unwillingness of the US to provide countercyclical lending—as manifested by monetary policy developments in the US causing a virtual halt of net US capital outflows—and the unwillingness of the US to provide an open market for goods, as manifested by the introduction of the Smooth-Hawley tariff bill in June 1930, triggered a global depression, the rise of extreme nationalism in Germany, which led to the Second World war.

6.3 What's Next?

The unhappy reality is that there are no guarantees that the driving forces of globalization are conducive to a different civilization based on cooperation, mutual respect and non violence. We are thus living in a period in which the rapid scientific and technological progress and free trade have paved the way to a global economy. However, political developments in major countries, the aggressiveness

---

[4] See Sommariva and Tullio (1987) and De Cecco 1973.

in the search for control of natural resources, the geopolitical instabilities arising from the consequences of climate change, and the exclusion of large parts of the world population from the benefits of the globalization processes could be at the origin of violent clashes in the future, pushing further away the establishment of a different civilization. Faced with these uncertainties, it seems only natural that the level of military expenditures worldwide remains high or increasing.

Our analysis indicates that progress towards an international system of independent nations based on cooperation and non violence depends crucially on consensus of interests and values, and on justice. Hereafter, we will analyze, without the pretence of completeness, the state of these consensuses.

Consensus on interests among the United States and its major creditors, particularly China, seem to be there. In the longer term, this consensus will continue if China changes its export led growth model towards a growth model based on internal demand rather than exports, and the Unites States reduces its large government deficits and debt, which is seen as a major obstacle to long term growth. Persistence of present policies could put these two countries on a collision course. Conservative group in both countries are already vociferous in blaming each and may gain more power in policy making if the present situation persists.

At the moment, the debate on how to reduce the large government deficits and debts in the United States is stalled by the evenly split electorate and by the obsession of the Republican Party on opposing tax increases, particularly to the wealthiest part of the population, and on pursuing deficit reductions through cut in government expenditures, particularly social programs. They sustain that inequality, arising from the present tax structure, does not matter as long as markets are working efficiently and economies are growing so that everyone is getting more.

There is some evidence that this is a purely ideological position. Their theory is correct in times when heavy industries and machines were the key to economic growth and a large proportion of rich people was required who could save their income and invest it in physical capital. However, there is a growing consensus among economists that technological advances have increased the importance of human capital in economic growth and hence the demand for highly educated workers. Education has now become the secret to growth. More equal societies are thus more likely to sustain longer term economic growth, since the more equal a society is the more people have access to education. How and if the Obama administration will able to tackle the twin problems of deficit and debt reduction and economic growth will be crucial for a peaceful evolution of international relationships in the longer term.

To an external observer, China's strategy to development since 1978 appear to rest on making sure that it provides political stability and prosperity. Since 1978, the export led growth model worked. China was mostly an agrarian society in 1978 and has become largely an urban, industrial society. A large number of Chinese have been lifted from poverty since 1978 and the economy has experienced unprecedented growth since then.

Since then, the Chinese regime has been active in building a constituency of support for this reform strategy, summarized in one sentence: devolving authority and yielding benefits. Devolving authority meant mostly promotion of the market and building institutions to supervise the market. Democratization was never under consideration. Despite some opposition to this strategy, there are reasons to believe that the large majority of Chinese people agree with this strategy, as they are happily involved in following Deng's advice: *It is glorious to be rich.*

During this period, China has become a full member of the international community. It has become a full player in the UN Security Council and an active participant in the WTO. Despite some rhetoric of sable rattling, mostly in the case of Taiwan, China has pursued an international policy leading her to seat at the tables were the international rules are written. It has not manifested aggressive foreign policies reminiscent of Japan of the 1930s and has not challenged the global power of the United States. This was consistent with her overall strategy of political stability and prosperity.

In 2008, the Chinese policy makers responded strongly to the global economic crisis. They launched an economic stimulus program that has largely offset the negative effects of the downturn in international trade. Although China's growth rates did slow down with respect to the past, the slowdown has been modest so far. However, some of the economic imbalances emerged in the early 2000s, combined with the slowdown of international trade, can pose a threat to prosperity. These imbalances are: the continuous reliance on investment and export to generate economic growth; the decline in households' disposable income as a percent of GDP; the relative low level of consumption in GDP; and the presence of an outsized manufacturing sector.

Several options exist to address these imbalances. Central to these options is the reform of the financial system. Over the past decade, negative real interest rates, the emergence of a significant informal credit market, the sharp decline of government bonds held by households are all indication of a repressed financial market. Liberalization of interest rates and the exchange rate are central to the reform of the financial system and the reduction of the saving-investment imbalances. These policy options, rebalancing the sources of economic growth in China, will help the United Sates in reducing its large fiscal deficit and put the trajectory of its government debt on a more sustainable path.

The question is if and how fast the new leadership in China will tackle the financial reform. Little is known on the economic policy views of the new political leadership. If past experience can guide us, it will take at least 1 or 2 years before the new leadership consolidate his power and put their mark on economic policy. It is thus too early to pass a definitive judgment on how the new political leadership will react to the global economic crisis. What one can say is that if the economic growth in developed countries remains sluggish or decline, this could affect economic growth and prosperity in China. As the legitimacy of the current regime in China rests mainly on the continuation of high growth rates, these developments could pose a threat to the present regime. How the regime will react to this perceived threat is difficult to say. What we can say is that these developments will reinforce the position of the conservative elements in the regime with a more aggressive nationalistic view.

Consensus on how to deal with the unregulated capital markets is less evident. The problems of unregulated capital markets became evident during the financial crisis of 2008. This crisis was the consequence of few facts described hereafter without pretence of completeness. Homeowners and other borrowers took up many loans to big to be repaid. Accounting frauds by many banks and financial institutions around the world made it possible to shift the risks involved with these loans to gullible customers and, at the same time, to generate large profits for these institutions. At the time of the financial crisis, as some of the banks and financial institutions had grown to big to fail, attempts to mitigate the breakdown of the system generated large amount of public money available for grabs.

At the same time, there was a large cry for introducing regulations of the international capital markets, as it was perceived that unregulated markets had failed to provide prosperity. Excess greed, excess concentration of power, excess lobbying and excess fraudulent activities were identified as the causes of this failure. The combination of our systemic perception biases and perverse incentives may motivate human kind to disregard the precepts of justice and favour ourselves at the expense of others. However, progress in how to deal with unregulated capital markets have been slow to materialize because of the power of the lobbies of the big banks and other financial institutions.

Consensus on justice both within and among our societies is also unclear. We have previously touched on this argument when discussing the debate on tax and equality in the United States. A similar discussion is now taking place in European and other developed countries. In Europe, the traditional welfare system built up during the last 16 years is crumbling as a consequence of the financial crisis and the necessary adjustments in fiscal policies. There is no doubt that the welfare state in Europe had gone too far in the past and has sapped the entrepreneurial forces necessary for economic growth. However, there is not yet a consensus on how far Europe should go in the reform of its welfare system without endangering economic growth. Lack of consensus may cause a crisis of the European Union, as Northern countries will grow far apart from Southern countries.

The debate on economic development and the redistribution of wealth and resources among developed and less developed nations has yet to reach a consensus. *First*, there are a growing number of people in developed countries who believes that international trade cannot benefit the poor, developing countries. They believe many trade agreements and multinational corporations can undermine the environment, labour rights, national sovereignty, and the third world. People who are not in favour of expanding international trade and desire preservation of local culture and customs are referred to as anti-globalists. Although these groups have no influence at present on policy making in developed countries, the persistence of sluggish growth and a widening gap between citizens of these countries may change all that.

The debate on wealth redistribution centres on the relative merits of international aid policies and the social, administrative and legal reforms in the poor countries. How to change international aid policies both bilaterally and multilaterally is the core of the debate. In the words of an African economist during a recent interview with a German journalist[5]:

> Huge bureaucracies are financed (with the aid money), corruption and complacency are promoted, Africans are taught to be beggars and not to be independent. In addition, development aid weakens the local markets everywhere and dampens the spirit of entrepreneurship that we so desperately need. As absurd as it may sound: development aid is one of the reasons for Africa's problems. If the West were to cancel these payments, normal Africans wouldn't even notice. Only the functionaries would be hard hit. This is why they maintain that the world would stop turning without this development aid.

In the eyes of the African, the large international development aid programs, channelled through the bureaucracies of the new African states, have resulted in corruption. Political elites of the African countries have frequently connived with transnational corporations in the exploitation of the rich mineral resources of the continent, leaving very little value added to the African nations.

Consensus on values is not yet there. There are different arguments here. *First*, cross cultural studies indicate that cultures that lavish physical affection on infants tend to be disinclined to violence. Also, societies where sexual activities in adolescents are not repressed seem to be less prone to violence. These studies indicate that cultures with a predisposition to violence are composed of individuals who have been deprived of the pleasures of the body during infancy and adolescence. Religious taboos, discrimination towards women, and militarization of societies are present in various segments of our societies and are among the factors that may have contributed to the predominance of violent behaviour of many peoples. Although we do not understand human behaviours well enough to be sure of the mechanisms underlying these relationships, these correlations are significant.

Second, there is an ongoing debate of the effects of globalization on diversity of cultures. Cultural globalization refers to the transmission of ideas, meanings and values across national borders. A vast group of people, extending from some religious fundamentalists to anti-globalization activists, tends to interpret globalization as a seamless extension of western cultural imperialism. They view the impact of globalization in the cultural sphere in a pessimistic light. They associate it with the destruction of cultural identities, victims of the accelerating encroachment of a homogenized, westernized, consumer culture. They often react violently in protest against globalization.

Another perspective, regards cultural globalization as a process of hybridization on which cultural mixture and adaptation continuously transform and renew cultural forms. The proponents of this perspective maintain that when two cultures trade with each other they tend to expand the opportunities available to individuals artists and intellectuals. The blossoming of world literature, the bookstore, the printing press, the advent of cinema around the globe are all cases in which trade has made different countries more creative, thus giving the world more diversity.

---

[5] See Thielke 2005.

6.4 Conclusions

The above analysis, without pretence of completeness, indicates that the basic elements for a stable world order among independent nations are not yet there. This is not new. Many of today problems, although in a different historical context, were experienced in the 19th century in conjunction with the rapid integration of the world economy and the industrialization of the agrarian societies in continental Europe. After the 1950s, the decolonization and the industrialization of essentially agrarian societies with different cultural backgrounds and the reintegration of the world economy are causing again tensions in the international system. The difference is that now we have the knowledge of the past and the intellectual means to analyze them.

In the present historical context, it is worthwhile to ask again the same questions that the moral philosophers posed: (1) what is the role of science in transforming our societies; (2) what is the meaning of freedom and democracy in a world dominated by few powerful political entities competing for finite resources; (3) what is the role of a private realm of society in shaping a less confrontational world; (4) how and why do human beings learn to cooperate, and what makes it for a constructive and useful life as opposed to a destructive one; and (5) what is the contractual basis of rights that would lead to tolerance, both in public and private spheres, and to the organization of nations in self governing societies through democratic institutions.

This book has put forward an answer to the first question by examining how science and technology could provide a way to explore and colonize other worlds within the Milky Way galaxy. It indicated that any space mission that involves transferring of people will have a significant economic, commercial and cultural payload benefiting all humanity and not just a few. Although emigration of people from Earth will not solve Earth population problems, which have to be solved on Earth, it would make the solution of these problems easier since people will have a choice. By widening the horizon and future targets for the exploratory instinct of humanity, science and technology will prevent human civilization to fall into apathy and stagnation.

Answering the other questions is not within the scope of this book. However, it is the hope of the authors that this book will stimulate scientists and scholars of humanities around the world to come together in challenging conventional social, political, economic and religious wisdom. Scholars of humanities should test their ideas in terms of logical coherence, explanatory powers, and conformity with empirical facts. The central problem is how to extend ethical concern to ever widening groups of people. In the past, we have learnt to extend ethical concerns to our families; our tribes; and our nations. How to extend ethical concern to all humanity is the next challenge. At the same time, a realistic biology of the mind, advances in physics, information technology, genetics, neurobiology, engineering, and chemistry of the materials are all challenging the basic assumptions of who and what we are, and of what it means to be human, thereby contributing to our evolving qualitatively in mind and spirit.

Efforts to enlarge the discussion to the informed general public are necessary, so that the achievements and controversies do not remain marginal disputes of a quarrelsome mandarin class but affect the lives of everybody on this planet. Advancements in information technology will certainly facilitate the global access to information. The coming together of humanities and science, as one culture, and the enlargement of the discussion to the informed general public may help in discovering ways to form a global society less confrontational and based on mutual respect and non violence. This is not a utopia, but a search for a value system rather than a set of shared beliefs, which should result in a world more varied, more interesting and more hopeful that may lead humanity to reach the stars in a not so distant future.

References

Bull, H.: The Anarchical Society: A Study of Order in World Politics. Macmillan Press Ltd., London (1977)

Chitnis, A.: The Scottish Enlightenment: A Social History. Croon Helm, London (1976)

De Cecco, M.: Money and Empire: The International Gold Standard, 1890–1914. Basil Blackwell, London (1973)

Field, A.: A new interpretation of the onset of the great depression. J. Econ. Hist. **44**, 489–498 (1984)

Metzler A.H.: Monetary and Other Explanations of the Start of the Great Depression. Journal of Monetary Economics 2 (1976)

Sommariva, A., Tullio, G.: German Macroeconomic History 1880–1979: A Study of the Effects of Economic Policy on Inflation, Currency Depreciation and Growth. St's Martin Press, New York (1987)

Stewart, M.A. (ed.): Studies in the Philosophy of the Scottish Enlightenment. Oxford University Press, Oxford (1991)

Thielke, T.: For God Sake, Please Stop the Aid. http://www.spiegel.de/international/spiegel/spiegel-interview-with-african-economics-expert-for-god-s-sake-please-stop-the-aid-a-363663.html (2005). 4 July 2005